Springer Series on Environmental Management

Robert S. DeSanto, Series Editor

Springer Series on Environmental Management
Robert S. DeSanto, Series Editor

Disaster Planning: The Preservation of Life and Property
Harold D. Foster
1980/275 pp./48 illus./cloth
ISBN 0-387-90498-0

Air Pollution and Forests: Interactions between Air Contaminants and Forest Ecosystems
William H. Smith
1981/379 pp./60 illus./cloth
ISBN 0-387-90501-4

Natural Hazard Risk Assessment and Public Policy: Anticipating the Unexpected
William J. Petak
Arthur A. Atkisson
1982/489 pp./89 illus./cloth
ISBN 0-387-90645-2

Environmental Effects of Off-Road Vehicles: Impacts and Management in Arid Regions
R. H. Webb
H. G. Wilshire (Editors)
1983/560 pp./149 illus./cloth
ISBN 0-387-90737-8

Global Fisheries: Perspectives for the '80s
B. J. Rothschild (Editor)
1983/approx. 224 pp./11 illus./cloth
ISBN 0-387-90772-6

Heavy Metals in Natural Waters: Applied Monitoring and Impact Assessment
James W. Moore
S. Ramamoorthy
1984/256 pp./48 illus./cloth
ISBN 0-387-90885-4

Organic Chemicals in Natural Waters: Applied Monitoring and Impact Assessment
James W. Moore
S. Ramamoorthy
1984/282 pp./81 illus./cloth
ISBN 0-387-96034-1

The Hudson River Ecosystem
Karin E. Limburg
Mary Ann Moran
William H. McDowell
1986/344 pp./44 illus./cloth
ISBN 0-387-96220-4

Human System Responses to Disaster: An Inventory of Sociological Findings
Thomas E. Drabek
1986/512 pp./cloth
ISBN 0-387-96323-5

The Changing Environment
James W. Moore
1986/256 pp./40 illus./cloth
ISBN 0-387-96314-6

Balancing the Needs of Water Use
James W. Moore
1988/280 pp./39 illus./cloth
ISBN 0-387-96709-5

The Professional Practice of Environmental Management
Robert S. Dorney
Lindsay Dorney (Editors)
1989/248 pp./23 illus./cloth
ISBN 0-387-96907-1

Landscape Ecology: Theory and Applications
(Student edition)
Zev Naveh
Arthur S. Lieberman
1990/384 pp./78 illus./pbk
ISBN 0-387-97169-6

Inorganic Contaminants of Surface Water: Research and Monitoring Priorities
James W. Moore
1991/360 pp./13 illus./cloth
ISBN 0-387-97281-1

Long-Term Consequences of Disasters: The Reconstruction of Friuli, Italy, In Its International Context, 1976-1988
Robert Geipel
1991/192 pp./81 illus./cloth
ISBN 0-387-97419-9

Robert Geipel

Long-Term Consequences of Disasters

The Reconstruction of Friuli, Italy, in Its International Context, 1976–1988

With 81 Illustrations

Springer-Verlag
New York Berlin Heidelberg London
Paris Tokyo Hong Kong Barcelona

Robert Geipel
Geographisches Institut
der Technischen Universität
Arcistrasse 21
Postfach 20 24 20
D-8000 München
FRG

Cover: Photo of a sculptured stone plate on the wall of a rebuilt house in Osoppo's main street. The seismogram with the tumbling letters of the year 1976 (upper frame) has been juxtaposed by the sculptor with the trunk of a broken-down tree. New shoots grow up toward the Virgin with the dove of peace in her right hand. The year 1982, when the owners moved back into their new home, is demonstrated also, but formed out of building blocks of rigid austerity. They might convey a feeling of seismic security but not any longer the feeble grace of the former arch.

Library of Congress Cataloging-in-Publication Data
Geipel, Robert.
Long-term consequences of disasters : the reconstruction of Fruili, Italy, in its international context, 1976-1988 / Robert Geipel.
p. cm. — (Springer series on environmental management)
Includes bibliographical references and index.
ISBN-13: 978-1-4612-7782-8 e-ISBN-13: 978-1-4612-3064-9
DOI: 10.1007/978-1-4612-3064-9
1. Earthquakes—Italy—Friuli. 2. Disaster relief—Italy—Fruili. 3. Regional planning—Italy—Fruili. I. Title. II. Series.
HV600 1976.F745 1991
363.3'495'094539—dc20 90-23219

Printed on acid-free paper.

Photocomposed copy prepared from the author's PageMaker file.

9 8 7 6 5 4 3 2 1

Series Preface

This series is dedicated to serving the growing community of scholars and practitioners concerned with the principles and applications of enviromental management. Each volume is a thorough treatment of a specific topic of importance for proper management practices. A fundamental objective of these books is to help the reader discern and implement man's stewardship of our environment and the world's renewable resources. For we must act to bring harmony to it, and nurture an environment that is both stable and productive.

These objectives have often eluded us because the pursuit of other individual and societal goals has diverted us from a course of living in balance with the environment. At times, therefore, the environmental manager may have to exert restrictive control, which is usually best applied to man, not nature. Attempts to alter or harness nature have often failed or backfired, as exemplified by the results or imprudent use of herbicides, fertilizers, water, and other agents.

Each book in this series will shed light on the fundamental and applied aspects of environmental management. It is hoped that each will help solve a practical and serious environmental problem.

Robert S. DeSanto
East Lyme, Connecticut

Preface

The assessment of long-term recovery of disaster areas, as Drabek points out (1986, p. 250), lacks an adequate base for drawing firm conclusions. The results offered in the literature rarely discuss the situation before the event, and often the measurement points for the recovery process are too close together in time. Investigators seldom have the patience to hold on to their topic for years, and often the composition of investigation teams changes between the beginning and the end of their enterprise. Moreover, disaster researchers are more inclined to collect data on various different catastrophes in order to compare them and to show "something new," than to cling to a single event for years and years. This could be interpreted as lack of flexibility or stubbornness, or even lack of ideas, and at least as a poor feeling for public relations.

In turn, critics of disaster research again and again deplore the lack of long-term studies. Such work is indeed difficult to organize. Funding institutions have to be convinced that it is worthwhile to invest in studies 10 years after a certain event. Public interest has often faded away in the meantime, and has been transferred to more recent catastrophes. The motivation of members of research teams (and the recruitment of graduate students to become members of such teams) is much more difficult than it would have been right after a disaster. To illustrate this with our research in the case of Friuli, in 1977, under the fresh impact of the event, six graduate students of our Geography Department volunteered to go for months to the disaster area in a most complicated situation. The aid of Friulian-speaking Italian students could easily be organized at Udine University, encouraged by a more or less symbolic payment for the interviewing of the victims. More than 6500 questionnaires could be collected among the stressed population of the disaster area. Ten years later, in 1986, the students recruited for a field trip, intended to train a new research team were reluctant to agree to work in the area. The destruction, poverty, and distress that had obviously motivated students to do such work in 1977 were lacking now. In the face of the sometimes sumptuous reconstruction that had since taken place, only two graduate students agreed to write their theses on topics related to our second research program. Italian students had to be paid heavily to make the interviews. "They live much better than we do," said the Munich students. "Why should I sacrifice one year of my academic career in order to do research here, if all problems seem to be solved?"

Altogether, five volumes of our departmental series of publications (Münchener Geographische Hefte nos. 40, 43, 45, 59, and 65) were dedicated to the Friuli disaster. The editors were inclined at last to ask themselves if there should not be more variation in their publications. Had not everything been said already on 1095 pages?

In the view of hazard theory, however, such stubbornness, as that which determined us to continue, has its advantages. A firm system of cooperation with local authorities, established years ago, granted access to otherwise inaccessible, unpublished documents and administrative data. Individuals interviewed long ago under the fresh impact of the disaster remembered us, and 10 years later asked us into their homes, proud of what they had achieved in the meantime, but also frank to discuss shortcomings of the reconstruction process. An atmosphere of trust and credibility ruled the second phase of our investigations. Photographs taken by our first team of the victims standing amid the ruins of their homes were proudly confronted with their newly built houses. One member of the first team, then a freshman in our activities, had finished his dissertation (Stagl, 1986) on man-made hazards and came back after additional training in Berlin and Boulder. He joined the second team as assistant director and could thereby add his experiences and good command of Italian to the new crew. Mrs. Mirella Loda, an Italian postgraduate from the University of Florence (Loda, 1990), who specialized in economic geography, joined too. Also, a new topic, "regionalism and sense of place," had been developed at our department in the meantime, and it proved worthwhile to combine this theme (with the help of another colleague from our department) with our task in Friuli, since the disaster had proven to be a starting point for new local identities. So again a team of six researchers could be formed. Their findings are reported in the following text.

Acknowledgments

The research project whose results are described here has benefited from funding by the German National Science Foundation (DFG) and from the advice and assistance of numerous people. Chief among these are the co-authors of a German preliminary version of this book, published in 1988, Juergen Pohl and Rudi Stagl, the research assistants Anna Bardola and Holger Hochguertel, our Italian partner from the Emergency Commissioner's Office Emanuele Chiavola, and Mirella Loda.

Since the author's training in English was originally based more on Shakespeare and Geoffrey Chaucer than on the present Anglophone scientific lingo, Philip Wagner and Les Heathcote undertook the heavy task to look into my translation and render complex Teutonic-English sentence structures into not much less intricate "real" English. The Geography Department of Flinders University in Adelaide, South Australia, and its head, Murray McCaskill, provided an atmosphere of collegiality and friendship, a well-stacked library, and the possibility to discuss critically our results.

While the author spent his sabbatical working on this book in Australia, back home in Munich Mrs. Gertrud Moessinger and Mrs. Renate Dietl were conscientiously typing the manuscript and assembling on their computers corrections drifting in from such far away places as Vancouver (Philip Wagner) and Adelaide (Les Heathcote), while the cartographers of the Geography Department of the University of Technology in Munich, Christian Elsner and Lothar Meier, showed their craftmanship in drawing the maps and figures.

I greatly appreciated the advice of Robert S. DeSanto, who brought parts of the developing manuscript, operating from East Lyme, Connecticut, to the attendance of Tom Drabek, hiding away in the Rockies high above Denver, and of David Alexander, divided between Amherst and fieldwork in Italy. The gratefully accepted improvements of this book are theirs, whereas any errors remaining are the responsibility of the author.

A word of thanks is due also to my colleagues in Friuli, Professor Giorgio Valussi, Trieste, and Professors Giovanna Meneghel and Raimondo Strassoldo, Udine, and to the people of Friuli, from decision makers in the town halls of its reconstructed cities to the small villages and its mountain peasantry, for their spontaneous good will and helpfulness.

Contents

Figures

Tables

Photos

I
Introduction

1. Topics of Long–Term Studies in Hazard Research

At first, of course, it is necessary to recall that on May 6, 1976 in the corner where Italy, Austria, and Yugoslavia meet, there occurred an earthquake of Rm=6.5 magnitude that killed 1000 individuals, wounded 2500, and made 80,000 homeless. Four thousand eight hundred square kilometers of Friuli were affected, comprising the mountainous and hilly parts of the provinces of Udine and Pordenone. During the first period of reconstruction, on September 15, 1976, a second earthquake measuring Rm=6.1 struck, totally demoralizing the population and thwarting the reconstruction attempts. Therefore, 32,000 people had to be evacuated to the tourist hotels of the Adriatic coast, which were unused in winter and mostly lacked heating facilities. Evacuation had already been used in the 1974 Christmas Eve disaster of Darwin (Australia) by Cyclone Tracy; we will come back to the strategy of evacuation later.

For the 33,000 people who remained in Friuli's disaster area, as well as for the later use by the evacuees, a frantic construction of prefab towns began. By the beginning of the summer season on the coast, just before the return of the evacuees in May 1977, 65,000 persons had found shelter in such prefab buildings, which varied greatly in quality (Davis, 1978). Permanent reconstruction of the towns and villages in due time allowed the resettlement of the evacuated population into new homes and the dismantling of the prefabs. This process was finished about 10 years after the disaster, in 1986/1987, confirming the principle that a successful reconstruction should be finished within 10 years' time.

Earthquakes, the grandest of natural hazards, belong to those systemic interactions between nature and man that exert the most massive effects and are least subject to human control. The geological disturbances affect the physical existence of people, as well as their socioeconomic organization over so long a period, that countermeasures undertaken must from the beginning be closely adapted to new conditions created by the earthquake. People's trust in hitherto untroubled relations to nature, in the places they lived in, and in an environment they felt safe in, is suddenly shattered. It is not only the earth itself that has been shaken.

In time, however, the manner in which individuals and society first reacted undergoes changes. Gradually in the process of coping with the effects of the earthquake, people recover their initiative. Reconstruction takes place under a sort of tension between the inclination to restore as it was before the catastrophe, that is, to unmake the tragic event as far as possible, and the endeavor to protect the human subsystem as much as possible from the consequences of a new earthquake, by making the spatial structure less susceptible to future disasters (for the case of Italy see Alexander, 1986).

In the face of the many problems of reconstruction, one effect that is often overlooked is the change of social and regional structure produced by an earthquake. Those aspects, which go beyond immediate crisis management, can only be grasped by long-term studies. Our follow-up study "Friuli 10 Years After the Disaster" will therefore investigate how people and society regained the initiative, and what effects the decisions in various phases had. First, we wanted to learn through our work about the consequences of certain choices made. It is the question of "What would have happened if. . ." that makes Friuli valuable as a lesson for the future. There will always be more disasters. Which "Do's and Don'ts" can we deduce from Friuli (Chapter VII)? Each strategy one chooses to overcome a disaster has its consequences, and many things could obviously have been done differently during reconstruction. What were the decisive turning points (Fig. IV.6)?

Which among current conditions of development can be traced back to pre-earthquake trends, and which to the take-off after the disaster? Is it possible to discriminate between them?

But we had other reasons, too, for taking up studies on Friuli again 10 years after the 1976 disaster. To learn from its reconstruction in particular means to discover whether developments were condensed into a 10-year time span that otherwise might have taken a whole generation of decision making—10 years of frantic reconstruction that achieved much but also destroyed much. By observing the destruction and reconstruction of 4800 square kilometers of fully developed regional structure we may perhaps comprehend general trends affecting peripheral mountain areas within Europe, as if we were looking through a time accelerator or time–lapse camera. Notwithstanding the uniqueness of the event and the individuality of the region and population, it is not implausible to suppose that in Friuli within 10 years things happened that might also take place within the next 25 to 50 years in other Alpine areas. For example:

- the abandonment of high–altitude agriculture
- the gradual abandonment of small, remote settlements in the mountains as permanent places of dwelling in favor of more centrally located bigger communes
- the use of remote settlements as leisure–time residences (Kariel and

Kariel, 1982)

- the concentration of work places in industrial zones of good accessibility
- the concentration of growth processes along the traffic arteries
- the sporadic increase of residential land use, dispersing over the scarce level land in the valley
- the increase of over–aged population in the mountains.

These effects have possibly been caused or reinforced by decisions made after the earthquake. Friuli can be considered as a laboratory, in which through a set of measures, laws, and incentives the state more or less intentionally took actions with a spatial impact. The resulting effects may possibly teach us about policies concerning spatial order ("Raumordnungspolitik") in other Alpine or mountain regions elsewhere; for instance, when we study the effects of a ubiquitous promotion of housing facilities or a concentration of jobs in industrial parks.

The frequently quoted "10 years after" (Kates and Pijawka, 1977) are therefore more than merely a round number or a motive to look back. "Ten years after" is oriented neither to the sequence of generations nor to economic cycles or electoral periods. Is it a mere numerical fetish, which because of the prevalence of the decimal system leads us to use periods of 10 years? In any case, it is the commonly accepted interval between major censuses, and in hazard research 10 years has always been taken as the deadline that, when met, betokens a successful reconstruction, and that when it is exceeded indicates that reconstruction has failed. How important these "10 years" are in the case of Friuli will be discussed later.

But have there not been other and bigger disasters, even in Italy, such as that of November 23, 1980 in Campania and Basilicata (Irpinia), which it might have been more urgent to research (Geipel, 1983) than the already well documented disaster of Friuli 1976? Why not Armenia 1988, San Francisco or Newcastle (New South Wales) 1989? One reason is certainly that the earthquake risk in Friuli is not "snow of 10 years ago," but ever present. The last tremor of Rm=4.1, VI Mercalli happened in Friuli only on February 1, 1988. In Italy, 15 million people live in outmoded buildings that would not survive a quake at VI magnitude Mercalli. Reconstructed Friuli did. And progress in hazard research, furthermore, as pointed out in the Preface, calls for follow-up studies, as some critics of previous work (Geipel, 1982) have pointed out. The next chapter will therefore explain what unsolved questions in hazard research suggested a return to the scene of events in Friuli.

2. Open Questions in Hazard Research

Hazard research developed in the United States as an interdisciplinary enterprise after World War II. Gilbert White's initial question in 1957 was why, despite all the governmental programs of the 1930s under Roosevelt's New Deal (TVA, etc.) and the $5 billion that had been spent since the passage of the Federal Flood Control Act of 1936, damages had continued to increase. The false security engendered by protection behind new high dams had led to growing investments in the floodplain. More people than before lived in valleys threatened by flooding.

Research of that kind later was extended to focus on human reactions to drought, earthquakes, hurricanes, volcanic eruptions, tsunamis, and so forth; and already for the 1970s yielded an annual damage estimate of about $40 billion worldwide. This is the "interest rate" human society has to pay for the use of risky environments. Burton et al. (1978) used for this context the equation: Net benefit derived from occupying a hazard zone = (total benefits derived from occupying the zone – net costs and losses from natural hazards – costs of adjustment to the hazards).

Society is willing to pay this rate because coasts, valleys, volcanic slopes, and climatically favored though seismically dangerous areas offer resources that can make it profitable to take these risks. Whereas during the 1960s and 1970s it was mostly the disasters themselves, their causes (detected through the natural sciences), their economic consequences, and increasingly also the aspect of forecasting that caught the interest of scholars, the 1980s concentrate more on *Community Recovery from a Major Natural Disaster* (Rubin et al., 1985). Overcoming the consequences of a hazard and comparing the strategies of reconstruction following a disaster becomes the focus of research, to draw conclusions from as many case studies as possible in order to find applicable regularities and strategies for dealing with future catastrophes. The necessary long-term studies, however, are rare. Research in the Anglo-American context has a short attention span and jumps from event to event. As long as the media dwell on a certain disaster, funds flow freely and research pays off.

Alongside individual case studies, more theoretically oriented comparisons attempt to take stock of the results of such studies. As early as 1977, Haas et al. edited *Reconstruction Following Disaster*, whose Major Insights chapter (pp. 261–293) was partly based on the four well-studied examples of San Francisco, Anchorage, Managua, and Rapid City, but mainly on a more intuitive interpretation of the few existing case studies, which mostly lacked exact data.

The cooperation of economists (Kunreuther, 1974) and especially psychologists added new aspects of research, and traditional results were seen under new viewpoints. Instead of asking "How do people react to disasters," a new intervening variable such as "perception" was introduced. The question was reformulated as "How do people perceive a hazard and how does this perception influence their behavior?"

The study of reactions toward uncertainty, comparison of hazards (Slovic et al., 1979), adaptation to risks from natural, or since the 1980s, man–made and technological hazards suffered from the stringing together of isolated studies that could only seldom be reduced to a common denominator. The empirical material flowed in abundance but the theoretical framework was only weakly developed.

Foster (1980), Friesma (1979), and Cuny (1983) therefore attempted, for the aspect of reconstruction, to filter out general experiences through comparison of long–lasting effects. Their endeavors were intended to lead to provisional theories on *Reconstruction Following Disaster*. The main difficulties in all their work revolved around either the empirical statistical data or the complexity of long-term reconstruction, so that their findings have only slight empirical support. None of these studies goes far beyond the findings of Haas et al. (1977).

Bolin's (1982) study on "Long-Term Family Recovery from Disaster" confines itself to households, specifically families, because the difficulties with data referring to whole municipalities in comparative studies are formidable. Rubin et. al.'s study of 1985 therefore attempts neatly to avoid this difficulty in order to attain a closer view.

Rubin et al. (1985) compare 14 separate cases from the United States of disasters that happened between 1980 and 1984, including hurricanes, blizzards, floods, tornadoes, and earthquakes. This study stressed ". . . local decision making during long-term recovery" (p. 10). A purely quantitive inventory such as that in Friesema et al. (1979: four communities, secondary statistical data, mostly economic overall community data) was rejected by Rubin because of the small amount of evidence yielded by such procedures. Fourteen separate cases were sampled instead following criteria of extent of damage, time, location, costs, and the willingness of decision makers to cooperate. Two–person teams stayed in the selected communes for about four days, interviewing the decision makers from the time of the disaster. During this time the interviews were made, data collected, and impressions noted. Three communities were visited a second time. Reports and impressions were sent back to the respective interviewees for comment and smaller interesting points followed up on letters or telephone. The main issue of the study was ". . . the role of key local persons in the recovery decision-making proces. . . " (p. 13).

The method was ". . . to let local citizens and officials determine whether they were satisfied with the recovery in their locality" (p. 13). But the method of exploring the role of key persons by questioning those very persons is one of the weak points of this study. And it is questionable also whether the time chosen for a qualitative measuring of recovery, about 12 months after the disaster, is not too early to obtain a conclusive assessment. The bigger the disaster, the more destructive the hazard, the less well can the recovery phase be assessed after only 12 months. The qualitative opinions about the recovery extracted from the interviews could hardly be combined with the quantitative data. These are the shortcomings of an otherwise commendable study.

Community recovery as a result of reconstruction, according to Rubin's qualitative method, is mainly influenced by the interplay of "A) personal leadership, B) ability to act, and C) knowledge of what to do (p. 18, Fig. II.1).

The "availability of resources (human and material)" (p. 7) is more or less personalized through these three factors, and reduced to the ability, ideas, decisiveness, and so forth of the main actors and their influence on less powerful decision makers. Whether other factors (e.g., amount of subsidies, prompt-ness of payment) played a role, and whether these "objective" situations beyond the influence of the communes facilitated the "ability to act, knowledge of what to do and personal leadership," Rubin could not find out. The amount of subsidies, however, is not merely a result of how well the mayor can somehow scrape up money, but is also based on a framework that might differ between the 14 communities under survey. It is surely true, however, that within this framework the three factors have a positive or negative influence on reconstruction. The technological side of natural hazards is dealt with in the journal of the International Society for Natural Hazards Studies, *Natural Hazards*.

In 1986 Drabek for the first time gave a comprehensive overview and a synopsis of all hazard literature up to 1985. He could rely on bibliographies from 1981 through 1984, the journals *Mass Emergencies*, the *International Journal of Mass Emergencies and Disasters*, and *Disaster* (p. 8), as well as on informal information from participants in the 1982 World Congress of Sociology, Disaster Research Section. Non-English literature (in French, German, and Italian) was made available to Drabek by translations. The material acquired through this procedure could be ". . . judged to be representative of the sociological knowledge based on human response to disaster extent in 1985" (p. 9). Stress is laid on anglophone sociological literature. But work by geographers is well represented, whereas psychological studies (e.g., Fischhoff, Slovic, Lichtenstein) are less extensively documented (see articles in *Environment and Behavior*).

Nevertheless, this inventory of sociological findings goes far beyond the framework of normal bibliographies and exemplifies by its mere scope how complex hazard-related issues in respect to time and system level can be, and where the frontiers of research could be made out between 1973 and 1985 (Table I.1).

This overview shows that, as usual, the focal point of hazard research is still more or less centered on the event itself, but that the disaster phase of recovery has gained in interest. In the face of this shift in interest, geography as a spatial discipline, concentrating on description and explanation of spatial structures, attempts in this book to link hazard management with regional development. When we started as early as 1976 with our research in Friuli, we were well within Drabek's framework with the stress on "emergency" during that time. But the concept was from the beginning to be open for follow-up studies.

When the first publication appeared in 1977, one year after the disaster, we presented a survey of 6500 households in the 41 totally destroyed communes of the category "comuni disastrati." This book was translated into Italian in 1979, and in 1982 in an extended version into English. In 1979 Steuer wrote a study on the perception of the simultaneous hazards of rockfall, flood, and earthquake in Friuli, and Dobler in 1980 wrote on the chances of development after the disaster. Five theses from members of the Friuli team were published in an abbreviated form in a German-Italian edition (1980), and Stagl's study of 1981 on the opinions of the planners of the 41 destroyed communes on reconstruction finished the first cycle of papers.

To use Drabek's terminology, the studies of this first cycle laid stress on mitigation (perception, adjustment), response (evacuation, emergency), and planning. All these studies gave a sound data base for further comparisons, when the German National Science Foundation (DFG) granted the funds for a second cycle of studies from 1986 to 1988. Here stress was laid on restoration, reconstruction, and planning (not in the sense of Drabek as a measure before the disaster but in its connection with planned reconstruction) (Table I.2).

Drabek's Table 10.2 (p. 409) on the focus of international research provides us with a framework for the classification of our own work. This table shows that the sectors on which our main interest was focused, in respect to their topics and major conclusions, were among those of low coverage, but overrepresented in our endeavors. The project "Friuli 2" aimed intentionally to correct for such deficiencies.

3. Specific and Broadly Representative Elements in the Reconstruction of Friuli

Our studies, because of their two periods (1976–1980/1985–1988), may possibly meet the prescriptions for hazard research that were outlined in the previous chapter. They can also avoid difficulties encountered by many other studies. Friesema et al. (1979, p. 23) point out that going back to a historic disaster that took place too long ago seriously reduces access to data (especially data compatible with the standards of the current state of the art in hazard research), whereas too short a time depth of a postdisaster study does not give enough time to assess consequences fully. A review of the relevant literature suggests, however, that the example of Friuli, with its combination of short–, middle–, and long–term observations from various disciplines (sociology: Strassoldo, Cattarinussi, 1978; economy: Fabbro, 1983; regional planning: Di Sopra, 1986; disaster-management: Chiavola, 1985) in connection with our own findings may well belong to the few cases where the corresponding limitations on the usefulness of research findings could be overcome.

We must contradict the expectation that a high degree of similarity in the

Table I.1. Coded results of hazard research in comparison between 1973 and 1985. (DRABEK, 1986)

	System level																	
	Individual			Group			Organizational			Community			Society			International		
Disaster phase	1973	D^a	%	1973	D	%	1973	D	%	1973	D	%	1973	D	%	1973	D	%
Preparedness																		
A. Planning and mitigation[b]	33	120	8.7	11	−2	0.1	16	89	6.4	33	146	10.6	11	107	7.8	0	97	7.0
B. Warning	32	43	3.1	11	18	1.3	10	4	0.3	10	0	0.0	3	6	0.4	0	1	0.1
Response																		
C. Evacuation	30	51	3.7	14	29	2.1	10	1	0.1	15	23	1.7	4	2	0.1	0	5	0.4
D. Emergency	68	68	4.9	35	8	0.6	53	69	5.0	50	65	4.7	6	1	0.1	0	11	0.8
Recovery																		
E. Restoration	33	13	0.9	18	7	0.5	29	15	1.1	33	57	4.1	11	−1	0.1	0	25	1.8
F. Reconstruction	16	98	7.1	6	47	3.4	6	36	2.6	14	46	3.3	6	16	1.2	0	53	3.8

[a] D = The difference between the number reported for 1973 and the number listed in Table 10-1, i.e., the 1985 inventory.

[b] Due to the format and emphasis of our earlier review, I had to combine the findings pertaining to planning with those included in mitigation, i.e., hazard perceptions and adjustments. The discrepant increase is due to the narrower focus of the earlier inventory which was limited primarily to materials located at the Disaster Research Center, Ohio State University. Also, two other qualifications should be noted. First, the number used for the 1985 base refers only to those listed in the text; thus it is slightly lower due to duplications that are referred to, but not quoted. Because of the revision process, it was not possible to reconstruct the actual number of coded findings. Second, a small number of findings selected for inclusion in the 1985 inventory appeared in the 1973 inventory. Rather than exclude these in the count, I included them so as to compensate for the underestimate just noted.

political and cultural systems within a nation state should diminish divergencies in the reaction to disasters (Drabek, 1986, p. 65). Hazard research must emphatically respect the singularity of every event, even when it takes place in the same state as others (Italy: Val Belice 1968; Friuli 1976; Irpinia 1980) and under the management of the same hazard manager (in the last two mentioned cases through the former Minister of Civil Defense Zamberletti).

Mitchell et al. made this point also in developing a "Contextual Model of Natural Hazards" (1989) claiming "that a natural hazard is strongly modified by environmental, sociocultural, economic, and political contexts in which it occurs."

In regard to its spatial scope, too, the Friulian case produced results going beyond those of previous studies. Most of the earthquakes described in the literature took place in relatively homogeneous areas, mostly in mountains. "The 'typical' event has the main, concentrated pockets of greatest destruction and loss of life in mountain foot or foothill areas. . ." (Hewitt, 1982, p. 27).

In Skopje (Yugoslavia, 1963) and Guatemala City (1976), destruction was confined to intramountain basins. Friuli belongs to the relatively few cases in which all topographic units, including high mountain valleys, foothill areas, the moraine amphitheater, and the plains were involved. We therefore can compare the response to the same disaster in different natural environments.

Friuli is instructive also in respect to the flexibility of disaster management. Planning under the condition of a disaster is always different from "normal" planning, but not all disaster planning is alike: ". . . disasters differ from more routine emergencies in at least six respects: (1) uncertainty, (2) urgency, (3) the development of an emergency consensus, (4) expansion of the citizens' role, (5) convergence, and (6) de-emphasis of contractual and impersonal relationships" (Drabek, 1986, pp. 46–47). This classification of characteristics of disasters has also been dealt with in extenso by Harald Foster's 1980 *Disaster Planning*.

Plans in the view of planners are fixed programs, whereas a "realistic disaster planning requires that plans be adjusted to people and not that people be forced to adjust to plans" (Quarantelli, 1982, p. 2). Planning, in fact, especially after disasters, is an endless process (Quarantelli, 1982, p. 2).

This endless process shows up in Friuli in a great flexibility of planning of both the regional and national programs. This is true also for the willingness of the administration to alter laws if necessary, to adjust compensations to losses for inflation, or to combat overheated building prices by importing construction firms from outside of the region and even from neighboring countries. The nation state Italy, acting as disaster manager, guaranteed to two provinces, Udine and Pordenone, an absolute equality in treating all persons concerned, and in this sense extinguished all individuality. There are also findings in hazard research, however, that show that some communities may recover from a disaster much more quickly than others. Two factors seem to be important in this respect: "(1) Characteristics of the community and (2)

Table I.2 Inventory summary: Percent by total.[a]

	System level																	
	Individual			Group			Organizational			Community			Society			International		
Disaster phase	T	MC	SF	T	MC	SF	T	MC	SF	T	MC	SF	T	MC	SF	T	MC	SF
Preparedness																		
A. Planning	2.6	1.5	0.7	2.6	0.8	0.0	4.6	4.1	2.7	5.2	5.9	6.2	2.0	1.1	1.0	0.7	0.8	0.6
B. Warning	3.3	2.7	4.4	2.0	2.1	1.0	1.3	0.7	0.7	1.3	0.4	0.6	1.3	0.5	0.4	0.7	0.1	0.0
Response																		
C. Evacuation	2.6	3.2	4.6	2.6	2.3	2.1	2.0	1.5	0.0	2.0	1.7	2.0	1.3	0.7	0.1	0.7	0.4	0.2
D. Emergency	2.0	4.5	8.2	1.3	1.3	2.6	3.9	4.4	7.1	3.9	5.7	5.8	0.7	0.8	0.1	1.3	0.8	0.4
Recovery																		
E. Restoration	1.3	2.1	2.4	2.6	3.1	0.2	2.6	2.4	2.1	4.6	4.4	4.6	1.3	1.1	0.2	2.0	1.6	1.0
F. Reconstruction	3.9	4.7	6.3	1.3	2.1	3.0	3.3	2.1	2.1	2.0	2.7	3.2	0.7	0.7	1.4	1.3	3.2	3.0
Mitigation																		
G. Perceptions	2.0	2.1	4.3	0.7	0.1	0.2	2.0	0.8	1.4	1.3	0.7	0.6	0.7	0.1	0.0	1.3	1.3	1.0
H. Adjustment	2.6	3.5	3.0	0.0	0.0	0.0	2.6	1.9	0.2	2.0	3.1	1.8	3.9	6.8	3.6	2.0	2.7	3.3

[a] T = Topic; MC = major conclusion; SF = specific finding.

characteristics of the event" (Drabek, 1986, p. 233).

Since the events in Friuli can be taken as having had a uniform effect throughout all places suffering the same degree of damage, the particular qualities and power of resistance of the various communities could be evaluated individually. We shall show this in this book with the example of four communities, all belonging to the category of "communi disastrati," in close neighborhood to each other but responding in different ways because predisaster socioeconomic conditions and personality types of the decision makers were different (Chapter V).

The tendency of Anglo–American researchers to treat things alike if they happen within one country must be rejected in the case of Italy. The preconditions for overcoming the crisis, or the initial situation of the disaster, show the special status of Friuli in comparison, for example, to the earthquakes of Val Belice (1968) in Sicily (Haas and Ayre, 1969), or Irpinia in the hinterland of Naples (1980) (Alexander, 1982).

1. Since it lies on the boundary between NATO and Yugoslavia, numerous army units with heavy equipment were concentrated in Friuli in the vicinity of the disaster area (e.g., in Gemona, where many soldiers died in their barracks).

2. Excellent traffic connections in the form of railway and freeways were found in Friuli because of its location in a three-country corner; they were mostly undestroyed and facilitated the transport of supplies during the disaster and shortly afterward.

3. The adjoining states of Austria and Yugoslavia, because of their related minorities living in Friuli, carefully watched all events. The reporting media, especially television, effectively brought the disaster to the knowledge of the Central European population and stimulated their willingness to help a region to which many persons had personal links because they were familiar with the sea resorts of the Adriatic coast.

4. The infrastructure of a European-class tourist area provided ample accommodations that facilitated the evacuation of the disaster victims. This evacuation (of more than 35,000 people) was politically enforced and the timing of the evacuees' departure, just before the tourist season started, was perfect. In the area of the Irpinia earthquake near Naples a similar planned evacuation was thwarted because of fears that this might lead to a "legalized permanent occupation."

5. The nearness of a practically unaffected big city (Udine), which could act as an operation center for rescue and reconstruction in its hinterland, was of great importance.

Because of these "five lucky constellations" (Chiavola, 1985), some of the paradigmatic relevance of the case study of Friuli for other situations must be retracted. Other exceptional features (differing from the hitherto experienced "Italian way of treating disasters") were the institution of 390 three-person teams, which within four months evaluated damage to 85,000 buildings, and the early start of legislative activities that granted funds for a first reconstruction.

The "Italian way of treating disasters" has in the meantime improved a lot because the administration has adapted itself in its way of operating to a whole trail of natural catastrophes since the basic law was passed in 1970, improved after the Friuli experience, and again reformulated as David Alexander shows in his 1986 study "Disaster Preparedness and the 1984 Earthquakes in Central Italy."

Atypical in Friuli, on the other hand, was the repetition of the earthquake after only four months, a fact that partly cancelled the positive effects of the "five lucky constellations" quoted.

4. Literature on Friuli and the Earthquakes in Friuli

Besides our own contributions during the first working period between 1976 and 1981 (Geipel; Dobler, 1980; Steuer, 1979; Progetto Friuli, and Stagl) and the English version of our first book (Geipel, 1982a), Friuli of course was also an object of interest for Italian scholars. In addition to contributions on geophysics, tectonics, reactions of buildings to seismic stress, and commentaries on the legal handling of affairs, rather soon politically inspired books were published (cf. Ronza, 1976; Turoldo, 1977). Many nostalgic contributions too, were dedicated to the lost works of art and architecture

Two geographers from the University of Udine, Barbina and Valussi, as early as 1977 pointed to shortcomings of reconstruction planning — although without exact figures — and doubted the rationality of further regional development. Especially important contributions came from Strassoldo and Catarinussi (1978). They dealt with the historic background of the area, regional structure and the character of the Friulians, the situation before the disaster, and the extent of international aid to the region. Empirical studies included a survey of 31 entrepreneurs with 434 employees in 19 enterprises located in 15 communes. A case study on Venzone with 80 interviews of victims deals with psychological and social effects of the disaster.

Also in 1978 a volume was published by Nimis, an architect who had already been planning for Gemona, Artegna, and Magnano before the disaster and could therefore give concrete advice for further developments and possible faults in reconstruction.

In 1977 a journal entitled *Ricostruire* started up, and until 1982 published numerous articles exclusively dealing with Friuli's reconstruction. Also, Hogg

(1980) published a case study on Venzone, which found (in contrary to us) the Kates/Pijawka model (1977) lacking.

In 1981, Cattarinussi et al. published a survey of 900 interviews in 16 communities, which were divided into three categories according to the numbers of people dwelling in prefabs in 1978. The socio-psychological effects of the earthquake on the population are described. In 1981 Cattarinussi and Pelanda also wrote on *Disaster and Human Action* in a more general way, basing themselves mostly on 1977/1978 data. In 1983 Di Sopra and Pelanda produced a *Theory of Vulnerability*. In 1985 Fabbro wrote on "Reconstruction of Friuli" based on 2200 interviews, focusing on regional, urban, and family effects and showing a profound knowledge of local details, the best book hitherto in the context of Italian research. In 1986 Fabbro also organized a conference on "Ten Years After" and published the congress papers. Mattioni's (1986) report on the industrial development was informative on the history of single enterprises but deficient in the random sample taken. Fogolini in 1987 reported on the perception of reconstruction by the citizens of Artegna, supplementary to our studies in this book on Gemona, Venzone, and Osoppo. In the same years Dynes et al. wrote on *Sociology of Disasters*, including Friuli into a wider framework. The last book that will be mentioned in this review is Loda's (1990) dissertation on "Earthquake, Reconstruction and Industrial Development. A Long-Term Study on Friuli" within the program of our Munich working group that will also be referred to in this book.

II
Overview of the Initial Conditions for Reconstruction

For an appropriate evaluation of any reconstruction it is necessary to go back to the initial conditions of an area, because the roots of success or failure reach far back into predisaster periods. Therefore it should be kept in mind that reconstruction cannot be evaluated on the basis of the mere physical appearance of an area alone (new houses occupied, firms flourishing, people in jobs, administration and infrastructure working well). What would be the status of the study area with and without the disaster? How would a disaster of a given extent affect an area with only slightly different characteristics (of history, people's mentality, socioeconomic development, location within the global/national/regional framework)? The example of Friuli may contain a lesson for the kind of hazard research that in searching for "laws" fails to take heed of individualities, especially in the context of a highly developed country like Italy with its cultural diversity.

1. Indemnification versus Insurance: Two Models of Settling Damages after Natural Disasters

Having in mind Mitchell's admonishing statements about a "Contextual Model of Natural Hazards," we should try to fix the position in which the case of Friuli must be seen as to the models of indemnification after disaster: the models of compensation versus insurance, because these models go back to two images of man.

If caused by man (war, ecological or technical catastrophe) people will rightfully claim compensation from the person or institution (government) or its delegate who is responsible for the disaster. Millions of individuals in Germany and elsewhere during World War II lost their homes because of bombing or they became fugitives and had to emigrate from home while others suffered no material losses other than a devaluation of their savings. The Federal Republic of Germany (FRG) as successor of the Reich raised funds for the 10 million refugees from its eastern provinces and from German–speaking areas of multinational states in Eastern Europe for a sharing of losses ("Lastenausgleich"), amounting between 1946 and the end of 1989 to about billion $80 billion (current exchange rate).

The predecessor state of the FRG, and not the German nation made up of its individual citizens the Reich and its political institutions, guilty of the suffering and extinction of millions of Jews in the holocaust, pay a compensation ("Wiedergutmachung") to the State of Israel and victims worldwide to the extent of about $49.4 billion.

"Lastenausgleich" and "Wiedergutmachung" were instruments of indemnification if the State is found guilty. But what, if not the State but Nature, in "Acts of God," causes disasters too great to be borne by the individual, its family or its political community? Who is "guilty" of an earthquake, flood, drought, or cyclone? Not government, indeed. But already discussion on the extent of the consequences of such a natural disaster brings many difficulties. The greatest losses of lives through an earthquake result from the collapse of buildings. Has the community, the state, or the national government issued building codes to safeguard against damages if the seismic risk of an area is known? Did it control strictly enough whether these codes have been followed correctly? Did it safeguard public buildings like schools, hospitals, nuclear plants, bridges, or dams through additional security standards? Is government allowed to accept the sports stadium of Berkeley (or its research reactor) right on top of the Hayward Fault? Is the Government of the Soviet Union allowed to build a nuclear plant in a potential earthquake region in Armenia?

If in Austria or Switzerland a local building commission would allow the construction of a building in an area threatened by avalanches, compensation would have to be paid in case of a disaster. In all of these cases the State would be an accomplice if not of the causes but of the size of the societal consequences of a natural disaster, be it by actions or neglect of actions. We are entitled to claim compensation from the State, and if it is a State founded on constitutional law we can cite it before court like any individual citizen because it has failed. But is this true also for medieval buildings in Gemona, constructed long before seismic safety rules were known and before the institution of present–day Italy?

Questions like these are being asked in order to generate awareness and sensibility for the complexity of the different solutions possible to compensate for losses of a disaster-stricken population. This awareness is not yet widespread.

Recent Developments Concerning the Insurance of National Hazards

In 1989/1990 three activities started at exactly the same time: the establishment of an International Decade for Natural Disaster Reduction (IDNDR) (Housner, 1989), the newly started Drought Subsidy Debate in Australia, and the reflections in the United States on insurance of earthquake risk just before the expected Big Quake in California (Brown and Gerhart, 1989).

Let us start with an example:

In Australia the subdivision of large pastoral stations (up to 400,000 ha) into smaller units has been a basic official land settlement intensification policy from the 1880s to the 1950s. In New South Wales, where there was a ceiling on the maximum size of such pastoral stations in single ownership, once officialdom recognized in the 1960s that such a size was economically inadequate under pastoralism in the semi-arid areas, selected owners were allowed to attempt "opportunistic cropping" to supplement their pastoral income. In the wetter period of the 1960s and 1970s some success was achieved but the 1982/1983 drought saw such crops withered and the soil eroded by dust storms (Allan and Heathcote, 1987). But what if now the seasons turn drier (even before the greenhouse effect begins to work), is not the government accomplice to agricultural drought, which it may not have generated in the physical sense but made more likely through its laws, encouragement, or toleration?

In Australia you may insure your property against earthquakes (as they are supposed to be rare events), but not against drought. Drought is an Act of God and losses may be partly compensated by the official disaster relief policies of the individual states and the Commonwealth. Between 1962/1963 and 1987/1988 58% of all Commonwealth disaster payments were going into drought losses, and in South Australia between 1977 and 1983 this share was 82%. This equalled subsidizing the areas with a risk of drought (Heathcote, 1990) on the account of "normal" regions. Only tropical Cyclone Tracy, which destroyed Darwin in 1974, the floods of 1974 in Eastern Australia, the bushfires of 1981 and 1983 (Healey et al., 1985), and the Newcastle earthquake of 1989 were the other main contenders for official disaster relief.

Legal and Political Contexts

Questions like these touch a complex that has been long since taken up by US-American economic research. As early as 1969 Dacy and Kunreuther opened the discussion with their book on the *Economics of Natural Disasters*. They claimed that big catastrophes always were the incentive for a reformulation of compensation strategies. The 1964 earthquake of Alaska (because of the special status of this isolated 49th state of the United States) made federal aid necessary more than elsewhere. But such federal aid was then, too, granted for the many hurricanes, tornadoes, and floods of the year 1964/1965. The unusual treatment of Alaska had opened the door for federal aid everywhere.

Also in the case of Darwin (Australia), it was the special status of the Northern Territory (under Commonwealth government) and its many civil servants in public housing that attracted intensive federal aid after Cyclone Tracy. But this might have been also partly so because of the emotions of a whole continent that the disaster had happened on Christmas Eve. Friuli, in turn, because of its location in a three-country corner and its "show-window" character toward Central European neighbors, and of autonomy movements here and in other parts of Italy, has challenged the central power in Rome to perform here with an especially high degree of efficiency. The cases of Alaska, Darwin, and Friuli make us reflect on the fact that great disasters, nested in time gaps of public memory, are especially effective in organizing new approaches to compensation policies.

The growing role of central government in compensating natural disasters is derived from the extensive lack of insurance for home or shop owners against many risks. In the United States neither the Federal Disaster Relief Act of 1974 nor the Robert T. Stafford Disaster Relief and Emergency Assistance Act of 1988 (although intended to improve the 1974 act) has taken provision for earthquake damage to private property (Brown and Gerhart, 1989, p. viii). Federal aid is intended only for necessary public accommodations. And since, as shown, legal regulation of public aid is always executed in emergency situations and therefore under pressure of time, should not this time such a regulation be considered *before* the Big Bang, and not after? If earthquake insurance has existed for three quarters of a century, but with nobody buying the policies, should not government either force the citizens to buy or require the insurance companies to include earthquake in their portfolios (Palm, 1990, p. 102)?

The former is not possible because of legal, the latter for economic reasons. Brown and Gerhart, however, propose a tying together of insurance and mortgage finance industries, provided possible antitrust implications are not against it. They would like to see government in the role of a great re-insurance company (which in fact it is, by paying compensation out of its budget, collected from all citizens of a state).

Deficits of insurance business are easy to understand: earthquake-, flood-, and hurricane–prone areas are well known. All the people not living in those areas therefore would refrain from buying policies and would deny to enter into a joint liability association of which they never would have a profit. In the threatened regions, in turn, all endangered citizens would need to be insured. The premium for the respective insurances would have to be exorbitantly high because coverage from payments from unthreatened areas would be lacking.

Countries in the Third World, out of sheer poverty, have to rely on international aid anyway (Stephens and Green, 1979). Highly developed countries, however, have the option to settle compensation after natural disasters by either direct governmental subsidies or by insurance.

In this second case we should also take into consideration the geographic, historic, and social context of these alternatives of "governmental compensation" versus "insurance" concepts.

Insurance or Compensation?

The countries from which we took our examples for the reason of comparison with Friuli were Australia and the United States. Both countries have a largely dispersed settlement structure (even in their capital cities). In the huge suburban or interurban residential areas of North America or Australia (with a proprietor's proportion of 70%) there are only few high–rise apartments for rent. The vast majority of housing is as bungalows. Settlement patterns as just described strengthen individualism and produce a sensitivity for security that is granted by individual home ownership — "My home is my castle" (Palm, 1990, p. 117). People living in regions of this type are more willing to take up insurance, at least because mortgage banks ask for proper securities for the money they have lent out. In Australia earthquake insurance is included in all house insurance automatically (in contrast to the United States), a fact made feasible by the relative rareness of the risk up to now. It took the 1989 Newcastle earthquake to bring home to insurance companies that this is not the case and that they will face difficulties now and in the future.

In the big cities of Europe, however, the proportion of renters by far outweighs that of owners and therefore it might be sufficient for many individuals to insure furniture and household goods only, but not the rooms in the house in which they live. Laws made for an urban Italy with a majority of renters have therefore automatically neglected the settlement pattern of a rural Friuli with a high proportion of home owners.

But also the central or federal status of states is influencing the behavioral style of its population. Does it wait for "the State" to act in situations of emergency or does it rely on smaller units (the community, county or shire, or region)? The federation or Commonwealth is active for its individual member states only in special cases and preferably for its weakest members (e.g., the "youngest" like Alaska, the "not yet" like the Northern Territory or parts of a Central State with secessional tendencies in danger of breaking away — like Friuli). An "Aid Berlin" program is feasible if emotions run high or if disparities are too visible, if poor (Saarland) and rich Laender (Baden–Wuerttemberg) are being compared. Otherwise, a federally organized state sees only small obligations and mostly refrains from direct action with the exception of cases of disaster (Palm, 1990, p. 39).

> Strong governments make weak citizens—
> weak governments make strong citizens.

Antitheses like these influence also the "culture of compensation" ruling different regions of the world in respect to the indemnification of disaster losses. In all poor countries, especially those of the Third World and in all countries ruled by State Socialism where it is difficult for citizens to obtain private property, it is certainly "the State"—if at all, besides international organizations—for whom the people look in expectancy—and too often in vain.

Therefore we restrict our comparison to fully developed countries of the rich Western world of capitalism like Italy, Australia, and the United States, of whom the first one is ruled in a strictly centralized manner, whereas the other two are ruled by a federal system.

The Individual and the State

The United States of America and Australia with their extraterritorial capitals of Washington, D.C., and Canberra ACT are examples of a widely accepted self–government of citizens' interest. The state has been put into the position of a nightwatchman ("Nachtwächterstaat") and it is far away, not even part of "my" Idaho, Maine, or Oregon. Only in wartime, in great economic crises (New Deal), or in disasters the State imposes on private life, a life that should in ideality be well insured against all cases of need. A Calvinistic culture of self–reliant existence prevails. Community life, self–organization of parents' groups, churches, political parties, and grassroots activities are much more important to the individual than the far–away "State" that is present only as the unpopular collector of taxes. Daily life is largely free of the State. There is not even an obligation to report. Society has few visible hierarchical elements. The neighborly "you" does not know a "Ye" to keep you at distance.

Let us set against this model of public behavior the activities set in a context of a Central State Authority, where literally everything—although now increasingly being questioned—is decided by "Rome." Every recruitment of a civil servant is executed by a "concorso" (competition) arranged in Rome, each student's loan, each appointment of an assistant professor is not decided on by a self–reliant faculty or a self–confident senate but the "Ministry" in Rome. The mayors, despite their flamboyant sash, are weak figures in comparison with the Segretario, a state-imposed administrator. Prefects (district managers) are appointed through Rome and not elected by the citizens of a county.

Because of all that the citizen leans heavily on the State, which he fears, and despises, because against a State that rules and demands everything, everything in turn is allowed. If you can profit from it, do it. The State is hierarchically segmented like the Church, very ceremonious in its offices and full of uncodified but firmly observed rules of behavior in dealing with the high and mighty.

If we claimed the "responsible and emancipated citizen" in the cases of the United States and Australia, we might use the term "subject" in the sense of a vassal for the second type of comportment in relation to the State. Heinrich Mann has written a novel on such a subject, depicting the Bismarck Reich. He called it *Der Untertan*. Because of proven obedience to the State, a "subject" may hope for the provision granted to him as a helpless victim. A citizen, in turn, has to look for himself.

Relevance to the Friuli Case Study

During the 14 years of our follow-up studies after the 1976 earthquake in Friuli we concluded that during the accomplishment of the disaster these two models of "subject" und "citizen" in respect to indemnification got confused. They overlapped insofar that compensation granted to vassals was received by emancipated citizens in full capability of minding their own business. The result was the exuberance of reconstruction.

The Italian State, convinced of the helplessness of its subjects in regard to natural disasters as proven by previous examples (Val Belice earthquake of 1968), in Friuli, its show window to the neighbors, went to extremes in respect to indemnification.

In turn, the Friulians, less helpless victims but emancipated citizens of a high work ethic, responded to this generosity by mobilizing private means (savings, self-help, solidarity of families, mortgage) to double the State's performance through their own activities. Whereas before the disaster one room was to the disposal of the individual, there are now two side by side: one constructed at the government's expense, another by oneself.

On top of all that it must be taken into consideration that out of political reasons the damage estimates at the beginning of the promulgation of the compensation laws were deliberately assumed too high. This is an experience, Dacy and Kunreuther claimed as early as 1969, offering some examples (1969, p. 9).

So, too, the estimate of damage of the earthquake of November 23, 1980 in Southern Italy (Campania and Basilicata) with $15.9 billion might be exaggerated. But even if the new laws of indemnification created in the case of Friuli would have been applied in full range to Southern Italy, their effects would have been smaller than in Friuli. These indemnifications would have gone to subjects (or even might have bypassed them), not to citizens who knew how to create a prolific return. Therefore legislators have to come to a clear understanding that indemnification laws— applied in the same country— might have quite different results in respect to reconstruction.

2. The Region and the Events of 1976

The Northeast Italian boundary region of Friuli–Venezia Giulia consists of four provinces: Udine, Pordenone, Gorizia, and Trieste. The port town of Trieste, although located at the utmost periphery of the region, is its capital. The biggest of the four provinces, Udine, was also the hardest hit. Within a distance of only 100 km this province covers the whole environmental range from high mountains (Mt. Coglians, 2780 m), foothills (epicenter of the earthquake on Mt. San Simeone, 1505 m), moraine amphitheater and alluvial fan of the Tagliamento glacier (San Daniele, 252 m), to the delta soils, coastal plain, and the lagoons of the Adriatic Sea (Lignano) and its famous resorts.

This area is also the farthest extension of the European Community (EC) toward the East, the EFTA country Austria and socialist Yugoslavia. Some of the economic developments of the North Italian plain have reached as far as Friuli. But only the growth axis from Pordenone via Udine, Gorizia, and Trieste received decisive impulses from this development, apart from the chain of booming sea resorts from Jesolo via Caorle, Bibione, and Lignano to Grado. North of the growth axis lies a region that has for centuries suffered emigration. Valusssi (1971) has described this phenomenon in detail. This is also the very region where the earthquake struck (Fig. II.1).

The tourist potential of this rainy mountain area because of the competition of the nearby Dolomites (a climber's paradise) and the Carinthian Lakes in Austria is but poorly developed. However, it bears the brunt of all the tourist traffic toward the sea resorts and suffers the extreme annoyance of a new freeway with only a few exits that might contribute to local income: a mere through transit without profit for the natives. Only Tolmezzo, as the central place of Carnia, has attracted commuters from the narrow valleys toward its industrial zone, and in that manner prevented even faster emigration.

The process of a rapid deagrarization diminished the percentage of agriculturally occupied persons from 40% in 1951 to 13% in 1971, and 6.5% in 1985. Besides emigration to other parts of Italy and abroad, medium–sized and small firms in branches like the furniture, mechanical, and textile industries afforded some jobs to poorly qualified workers. Emigrants waited to return to Friuli in case these conditions should improve. A take-off set in in the early 1970s, but then the earthquakes struck Friuli in May and September 1976.

Sixteen thousand tents, railway cars, small trailers, and the first prefabs gave shelter to the 85,000 homeless survivors. But only evacuation into the hotels of the Adriatic coast, empty and cold in winter and only constructed for the May–September season of middle-class mass tourism could save children and old people, 32,000 of whom found shelter here.

"To help improve disaster management and to guide better environmental stewardship in yet to be encountered disasters it is necessary to come to a better understanding of long and short term consequences of past disasters and the associated coping strategies" (DeSanto, 1990).

One of the strategies is *evacuation*. Evacuations in times of peace are necessary but as to their legal consequence they are controversially debated instruments of disaster management. They strongly rely on the psychological status of the afflicted population. The case of Friuli (1976) and the case of the Darwin tropical Cyclone Tracy of Christmas Eve (1974) are comparable in the role evacuation has played. Only two years apart, the numbers of the evacuees were very similar, with 32,000 in the case of Friuli and 35,000 in the case of Darwin (Stretton, 1976).

In both cases the decision makers were strong personalities: Under Secretary of the Interior Giuseppe Zamberletti for Friuli and Major General Alan Stretton (as Director–General of the Natural Disasters Organisation of Australia) in charge of the rescue operations for a totally destroyed city of 47,000 inhabitants without electricity, water, and shelter against the downpour of tropical rain and threatened by the outbreak of disease.

But while the Friulians were evacuated to the hotels and condominiums of the Adriatic sea resorts of Lignano, Grado, Iesolo, and Caorle (Fig. II.2) by fleets of trucks and buses over a distance of only 100 km, 25,600 citizens of Darwin had to be flown out by the biggest air lift in peace time to the capital cities of Australia, mainly Adelaide, Sydney, Melbourne, and Brisbane at least 2,400 and up to 3,500 km away from Darwin. Nearly 10,000 left their destroyed town by road, that would lead them eventually as far as Adelaide, 2,500 km away (Chamberlain et al., 1981).

The evacuees of Friuli lived within the infrastructure of a Central European high–grade tourist area, in giant hotel palaces or commandeered apartment houses that were normally housing thousands of well-to-do tourists in a few hot summer months but not equipped to shelter fugitives of a disaster in winter and therefore mostly without heating facilities.

The Darwin evacuees mostly found a shelter with their own relatives (61% out of a sample of 500 interviewees). Both evacuations were claimed as a success of disaster management and there is ample evidence that these claims were justified. The November to May evacuation gave Friuli's emergency commissioner the time necessary to set up 21,000 dwelling units with more than 800,000 m^2 and Darwin's citizens were safely out of the way while the big clean–up in Darwin removed not only the still dangerous debris of collapsed houses but also the contents of thousands of refrigerators, freezers, and stores of frozen food rotting in the tropical sun since electricity failed.

However, the instrument of evacuation must be balanced carefully against the possible counterproductive outcomes like looting or permanent occupation of commandeered shelters by the victims of a disaster.

Only four years after the Friuli earthquake, a much bigger one occurred in the northeastern hinterland of Naples in South Italy. Thirty five hundred people lost their lives, and about 300,000 were made homeless. Because of his success in Friuli, Emergency Commissioner Zamberletti was again in charge of all operations, now with even greater power since he had been promoted to State Minister of Civil Defence. He of course intended to use the strategy that had proved successful in Friuli: to commandeer hotels, condominiums, and second homes in coastal cities like Sorrento, Amalfi, Positano, Capri, or Ischia.

But in the vicinity of a city of more than a million inhabitants like Naples, which had 80,000 homeless even before the disaster, plans of evacuation had to fail. Evacuation in the context of Naples would have turned into the occu-

Fig. II.1

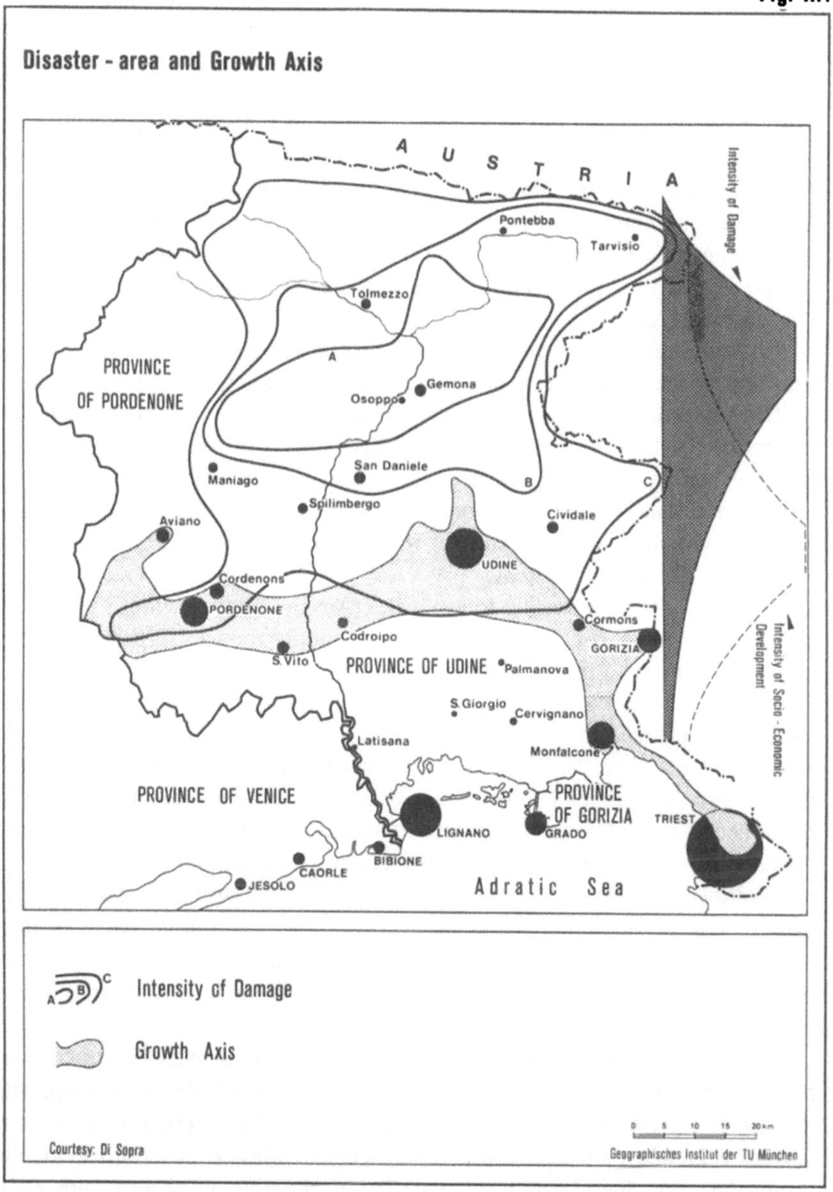
Disaster - area and Growth Axis
AUSTRIA
Intensity of Damage
Pontebba
Tarvisio
Tolmezzo
A
PROVINCE
OF PORDENONE
Gemona
Osoppo
San Daniele
B
C
Maniago
Spilimbergo
Cividale
Aviano
UDINE
Cordenons
PORDENONE
Cormons
GORIZIA
Codroipo
S.Vito
PROVINCE OF UDINE
Palmanova
Intensity of Socio - Economic Development
S.Giorgio
Cervignano
Latisana
Monfalcone
PROVINCE OF VENICE
PROVINCE
OF GORIZIA
TRIEST
LIGNANO
GRADO
BIBIONE
CAORLE
JESOLO
Adratic Sea
Intensity of Damage
Growth Axis
Courtesy: Di Sopra
0 5 10 15 20 km
Geographisches Institut der TU München

pation of private property forever, with the State as accomplice during and after the fact. Quantity made up for a new quality of problems. What seemed to be an excellent strategy in the cases of Friuli and Darwin failed in Naples.

Alexander postulates that the principal problem over transferring the survivors of the Irpinian earthquake was their reluctance to leave their home villages: the government had not taken into account the cohesiveness of local communities or how the local agricultural economy was to be sustained by transferring its workers out of the area. In the 1984 Abruzzo earthquakes, the same author found that the distribution of evacuees bifurcated: in those Abruzzi settlements in which the local administration wanted to draw the government's attention to their plight went for total evacuation, whereas the others evacuated in proportion to the number of houses damaged. Evacuation in Italy is easily politicized and not only a legal or practical measure in case of a disaster (Alexander, 1989).

The seemingly successful evacuation of Darwin also found a harsh critic. Members of the Departments of Anthropology, Sociology, and Social Work at the University of Queensland in Brisbane administered a questionnaire to 500 victims of Cyclone Tracy, of which 90 had stayed in Darwin, 107 were returned evacuees, and 219 nonreturned evacuees (the others were absentees during the disaster). The questionnaires intended to measure stress and found that those who had stayed in Darwin were evidently better off than the evacuees in a whole series of items.

Should it have been wrong to evacuate? Are the "stayers" those who made the "better choice," because they had not to endure the hardships of a strange new environment on top of the disruption of the familiar one by the cyclone? Comparing Western and Milne's (1979) results with our own I would claim that being sentenced to wait and do nothing produces higher stress than to stay and work hard. The "stayers" out from soaked crash housing in a tropical climate were indeed condemned to "hard labor." But they were also caught by the pioneers' mentality and the pathos of reconstruction. The evacuees, in turn, saw themselves in the role of beggars, claiming compassion.

A follow–up study of Western and Milne together with Chamberlain et al. showed that the higher stress of the evacuees was also due to fright of their house debris being looted, while the "stayers" experienced the solidarity of those who struggled side by side with them. Their feelings for Darwin were much more strongly engaged.

The evacuees experienced two forms of stress: the physical destruction of their home town plus the social destruction of their familiar environment because of the evacuation. Twenty five percent of Chamberlain et al.'s interviewees regretted their decision to be evacuated. Mass evacuations evidently thwart the great feeling to have mastered a situation in accord with ones' neighbors, to be turned from subjects to objects.

"That some scale of evacuation was needed is not being questioned, but the question which has to be asked is whether the indiscriminate method of

Fig. II.2
The Course of the Evacuation Process

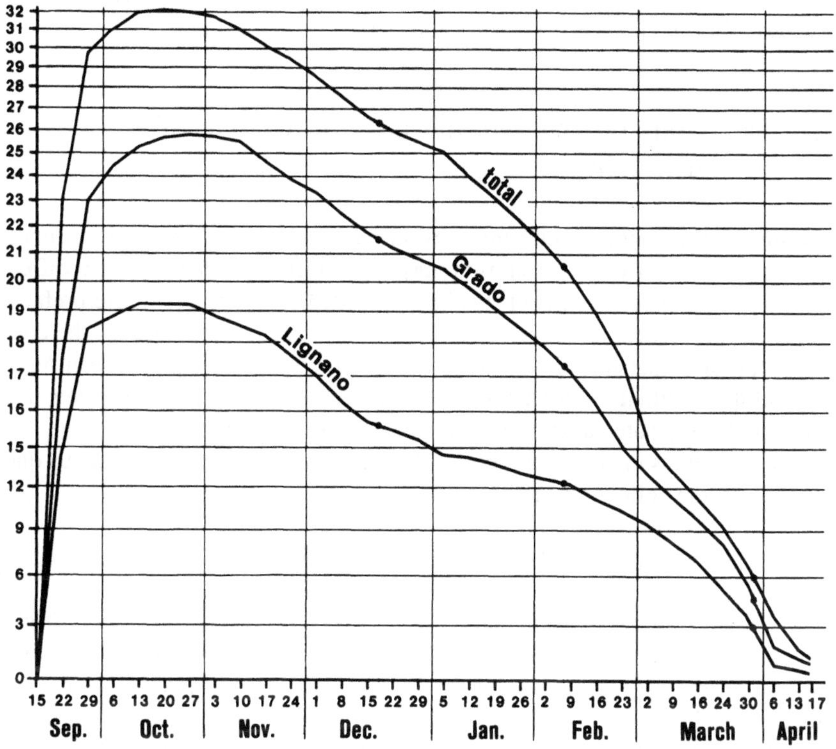

Courtesy: Geipel 1977, p. 76

evacuating residents vast distances with the sole aim of population reduction might not, in a future similar disaster, be modified by some form of selection" (p. 250).

"Finally, whether to evacuate natural hazard victims remains a problem. In Darwin, the evacuation was seen as urgent and necessary at the time, but since that time questions have been raised about its consequences. Short-term emergency needs and longer term consequences, both for individual well-being and the continuation of the community, need to be weighted against each other when decisions have to be made in future about evacuation. The weights on each side of the scale have not yet been fully uncovered" (Chamberlain et al., 1981, p. 154).

Disciplined people, as they are supposed to be, the Friulian evacuees left the hotels to return to the disaster area exactly in time to rescue the summer season in the Adriatic sea resorts.

They did not return to reconstructed houses but to 21,000 prefabs for the then 65,000 homeless. These "prefabbricati" were of varying quality and were to provide provisional homes for nearly a decade. We will discuss these "prefab-villages" later in detail (Chapter III).

3. Decisions on the Method and the Course of Reconstruction

Disaster management is confronted by many "bifurcations" of decision making, and whichever option is chosen may lead to irreversible courses in different directions. Switch points of this kind included the decisions to use prefabs, and about their types, their siting (dispersed or in prefab settlements), the concentration or dispersion of final reconstruction, and the choice between repetition of the traditional settlement pattern and creation of a completely new one that might anticipate future trends of residential development.

In Friuli, crisis management aimed at the rapid restoration of the "status quo ante," with the following options:

1. Earthquake of May 6, 1976

Tents as emergency measure with the policy "Dalle tende alle case" (right from the tents into the newbuilt houses) without the detour via prefabs, because these had proven to be permanent in the Val Belice earthquake of 1968. For this purpose law No. 17 was enacted on June 7, 1976 (just one month after the disaster!), which was to lead to the payment of subsidies to allow repairs to begin.

2. Earthquake of September 15, 1976

Alteration of priorities: no more new tents, which would have proven unfit for winter but (1) evacuation to the coast, (2) restoration of workplaces and public facilities of infrastructure, and (3) construction of prefabs.

As early as spring 1977 (April/May) "return to normality" can be attested within the prefab towns. The three objectives of a traditional society of "father at work, mother in the household, children in school," in however provisional a way, were implemented one year after the disaster (Fig. II.3):

What was the kind of political decision that had to be made beween Fall and Winter 1976 and Spring 1977?

One decisive question is in which form material losses should be compensated. Since there is no insurance against earthquakes in Italy direct government intervention takes place, as Alexander (1986) outlines in his chapter "Brief History of Earthquake Relief in Italy" in the study already mentioned. Franz (1979) distinguishes three kinds of basic social programs, among which one or a combination of several must be chosen for disaster management:

1. Restoration of the "social status quo," which means that the extent of compensation is a function of the individual loss. Whoever loses much will get a large amount in compensation, which leads to the preservation of social differences and the retrenchment of inequalities at least during the period of reconstruction.

2. The apportionment of aid follows the principle of equality, which means that every household (or person) gets the same material aid (quota per capita). The effects of this principle of equality lead to diminishing social differences (at least for some time).

3. Distribution of aid follows the principle of need. Subsidies are supplied only or mainly to those groups that cannot help themselves, totally or partly. This leads to a concentration of aid on the lower social classes.

a) With Law 30 on repairs, issued June 20, 1977, and Law 63 on new construction issued December 12, 1977, a mixture of the alternatives 1 and 2 of this list was made possible:

   • Close to alternative 1 lies the possibility for owners of several houses to receive subsidies for all of them (although with modifications, and to a smaller extent, for the second and all further ones)

   • Close to alternative 2 lies the possibility for an owner or renter of a house or dwelling, resident in an affected community at the time of the disaster, to receive a subsidy scaled to the size of the family, and expressed in square meters.

The *advantage* of this procedure was social equity. Furthermore, those who had been only renters before could become owners under these laws. The *disadvantage* of the procedure consisted in the fact that the value of the lost dwelling was not taken into account. Rumors say that "every pigsty was compensated," whereas young families were excluded from this procedure because they owned no title deed at the moment of the earthquake, having still lived with their families.

Fig. II.3

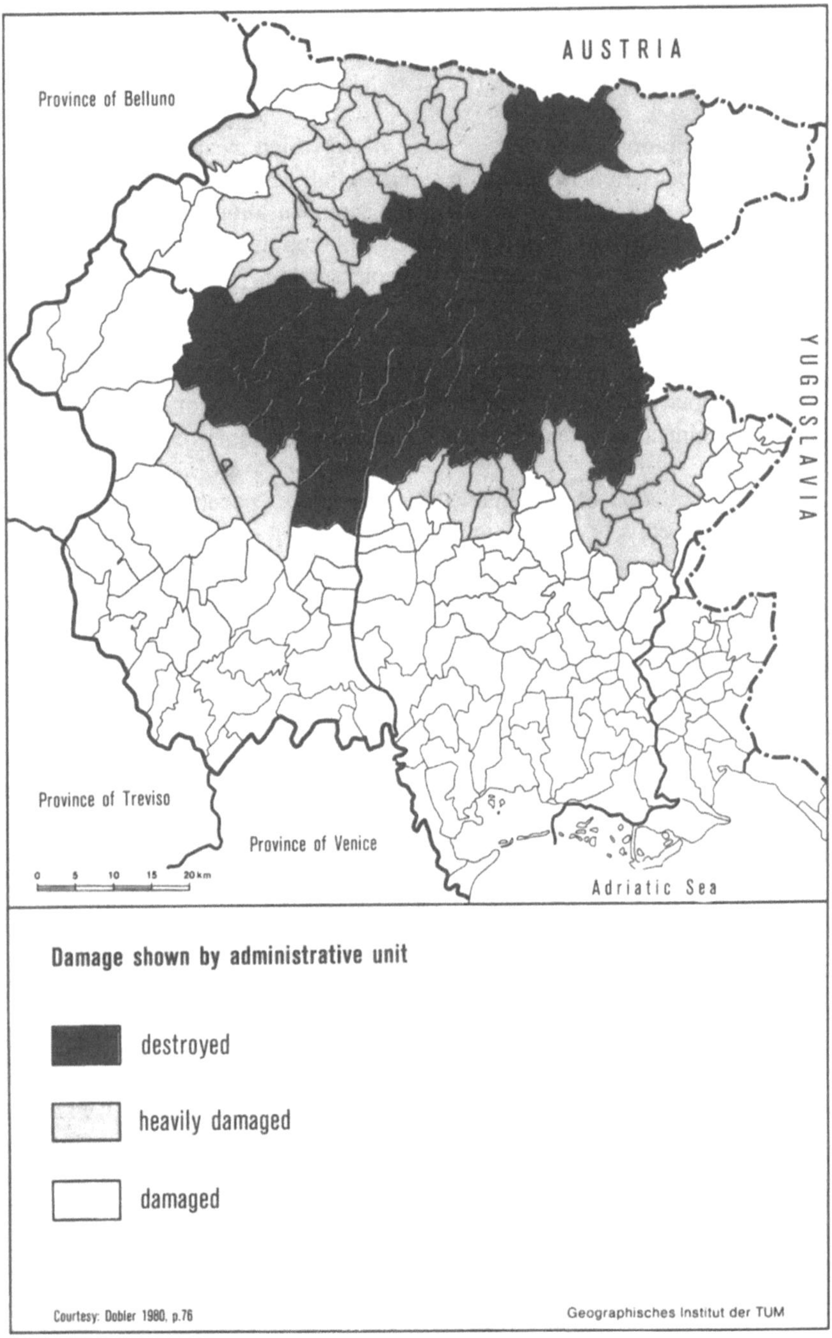
AUSTRIA
Province of Belluno
YUGOSLAVIA
Province of Treviso
Province of Venice
Adriatic Sea
0 5 10 15 20 km
Damage shown by administrative unit
destroyed
heavily damaged
damaged
Courtesy: Dobler 1980, p.76
Geographisches Institut der TUM

b) The second great issue centered on who should be responsible for building: private persons, cooperatives or the State (provinces, regions, communes). On a private basis 31,000 smaller and 29,400 larger repairs and 14,000 new constructions were completed. Cooperatives or public institutions carried out 12,600 major repairs and 1600 new constructions. Therefore in Friuli both forms coexisted side by side, together executing 89,000 projects.

c) A third issue concerned who should make the decisions on how and where to rebuild. Two ways seemed feasible:

- centralized, by region

- decentralized, through the community.

The decision was made between Summer and Fall of 1977 in favor of a decentralized procedure favoring decision making through the communities.

d) A fourth decision had to be made on the alternatives of either undertaking more repairs or practically building everything new. This was an issue that had much to do with regional identity and the conservation of local cultural distinctiveness. We will discuss later the case of Newcastle (New South Wales, Australia) where the earthquake of December 28 brought about a fierce controversy between developers and conservationists about demolition or reconstruction of heritage precincts in the CBD of this city. Beyond that, because of differences in costs in Friuli the question of repairing versus rebuilding was of economic importance. The ratio between totally destroyed and damaged buildings was 1:6, offering the options of:

- repairing buildings that were nearly totally destroyed, in order to restore the architectural predisaster situation, especially of buildings of high artistic value (such as cathedrals, medieval town halls, etc.) or

- building everything new, a decision that with the exception of the historic city centers (Centri Storici) of Gemona and Venzone was the rule in most of the communities.

Under the imminent impact of the disaster, the conservation of the cultural environment was pushed into the background. Toward the end of the building period, conservation of the cultural heritage gained impetus. The length of time that it took to decide between the alternatives of repair (Law 30:9 months after the May earthquake) and of total reconstruction (Law 63:15 months after the September earthquake), and the transfer of competence to the communities brought about a "return to normality," which meant the restoration of the spatial structure in a predisaster fashion, without anticipating a possibly more modern settlement pattern. Chances were lost to promote bigger communities with concentrated work places and better infrastructure. The egotism of the individual communities prevented the abandonment of the most remote and impracticable locations. The illusion that everything would be just as it had been before the disaster thwarted the creation of a regional structure that might have anticipated future trends of population, mobility, and lifestyle preferences, especially for the next generation.

The option for only one of three possible aims in Friuli raises the question of competition among three objectives normally encountered in postdisaster decision making:

1. the desire to return to normality as quickly as possible (the typical case, as many studies show)

2. the chance for a totally new start (e.g., concentration of all victims in one location: "Udine Nuovo" with 85,000 inhabitants)

3. the prevention of future catastrophes by avoiding the seismically unsafe area (a solution that has many backers at the very moment of the disaster, whose number decreases in the course of time in favor of option 1).

Of course, as in so many other cases, Friuli followed option 1, which meant as little planning as possible, therefore no loss of time and more "action" from the beginning, especially for active and potent people like the opinion leaders in many communities. Instead of long meetings and bargaining over the right way, the slogan "let the most able take over" was followed, especially by those who were capable of work and had to live in tents or later in prefabs and felt the drive to act independently from bureaucracy.

As Alexander points out in a personal communication, time is usually a necessary component of reconstruction, as planning must be followed by consultation and ratification. In a heavily damaged settlement of, say, 6000 inhabitants, any honest attempt to prepare a geological site plan, a reconstruction plan and a restoration plan for damaged buildings that are not to be replaced (and to commission building firms to do the work) would take at least two to four years. Such a plan prepared in 1984 for Sant Angelo dei Lombardi in Irpinia received 44 legitimate objections, so that decision making in Friuli seems rather exceptional.

Decision making by the communities also meant that an impatient population could put pressure on the communal authority charged with decision making, much more so than could have happened with a distant regional government. Competence for decisions within the communities closed out any different option from being discussed even in theory. Therefore, with the exception of man–made disasters like Seveso or Chernobyl, it is the rule that reconstruction rebuilds in the same place.

"The relocation of communities is extremely costly both in financial terms and in social upheaval. It cannot be considered as a viable solution to move whole communities, but it may be viable after disaster to move parts of overall settlements from dangerous zones" (Davis, 1981).

Thus, in Friuli, too, a community reorganization failed to occur, the plan of "Udine Nuovo" was abandoned very early, and only the Portis fraction of the community of Venzone was relocated in a new place because of the imminent danger of further rockfalls (Steuer, 1979).

The four phases of overcoming a disaster, emergency, restoration, reconstruction I, and reconstruction II, reflect the change during the course of time in the relation of man and environment as an interacting system (see Chapter IV). Our first book (Geipel, 1982a) gave a report on the first two phases. The following concentrates on the two other phases, and consequently starts with the leaving of the prefabs, the so–called "baraccopolises" of Friuli.

III
Effects of Provisional Housing in Prefabs: Slums of Hope or Despair

The establishment of makeshift housing for bridging the time between living in tents and moving into finally rebuilt homes is not necessary in all disasters. But mostly it is an unavoidable and expensive measure.

It became necessary after the second earthquake in September, 1976. If in May there still might have been some trust in the slogan "Dalle tende alle case" (right from the tents into new houses), it became clear after the second quake that evacuation and the construction of prefab towns would be unavoidable as intermediate steps to normality.

"Normality" in the opinion of the victims would mean returning to a new and "real" house. In that respect prefabs are exceptions and not a restoration of normality. Because of the limited stay in these prefabs they can be seen as a short-term adjustment. "Living in prefabs" was an important enough topic to deserve a special study as a master's thesis in our research program (Holger Hochguertel).

The use of prefabs may be an acceptable choice for the victims. But this measure also implies a whole series of problems; some of them will be discussed in the following paragraphs.

1. Expenditures

The decision to build prefabs after the second quake of September, 1976, for a projected time span of, say, five to seven years, was necessary in order to keep the population in its home area. The expenditures for this semipermanent reconstruction in the form of prefabs amounted to approximately 380 billion lire ($280 million at 1985 value) and their complete demolition will call for another 100 billion lire ($74 million). Since 750,000 m^2 of provisional living space were constructed, 1 m^2 of a prefab will have finally cost 650,000 lire, equivalent to $560. This is only a little less than 1 m^2 of permanent living space (plus a lot) would cost in this region of Italy. Davis (1977, p. 34) mentions that the igloos of Gediz, Turkey, were more expensive than the final permanent buildings.

Table III.1 Persons and families in prefabs

	5.77	3.78	9.79	10.81	10.85	6.86
Persons	65,438	61,219	43,000	38,859	25,838	20,535
Families	?	26,188		16,350	10,456	8,205
Dwellings	20,957	21,207		15,701	12,229	9,727
Persons per family	?	2,3		2,4	2,5	2,5
Persons per dwelling	3,1	2,9		2,5	2,1	2,1
Families per dwelling	?	1,2		1,0	0,9	0,8

Source: Ricostruire 10/11, 1980, p. 58.
Geipel, 1977; S.G.S., 1985, 1986; ISTAT, 1984.persons per

Alexander describes the failing of this policy in the 1968 Val Belice, Sicily, earthquake, as follows:

"Although some 14 laws governing reconstruction in the Belice Valley were eventually passed, and about 1,000 billion lire spent on public works, the net effect has not been encouraging. About 40,000 prefab units were constructed, incredibly, at an estimated 83% of the cost per square meter of full-scale reconstruction. Many of these flimsy dwellings were still in use when I visited the area 15 years after the disaster, while others had been abandoned, disfiguring the landscape. Political machinations and corruption are blamed for the absolute inefficiency of the reconstruction in Sicily (which has become a cause célèbre) and 33 legal cases have ensued over construction problems (Di Giovanna, 1974)" (Alexander, 1986, p. 25).

In turn, we are not astonished about the high costs of makeshift housing in the form of prefabs. Up to now, alas, they seem unavoidable. Not all of them were used. Davis explains some underutilization with the following reasoning:

"Reasons for this underuse may include: overestimates of the homeless population; excessive volumes of aid; the location of the units (often away from bus routes, a vital requirement as work gets back to normal); cultural rejection of unusual forms of housing; the almost universal hostility to multi-family units; and finally, the fact as more permanent housing becomes available, this is seen as a better alternative" (Nicaragua, Davis, 1977, p. 32, following Drabek, 1986, p. 246f.).

Interesting observations on the longevity of prefabs were also given in the

1979 issue of *Sapere* dedicated to Friuli and in *Solbiati and Marcellini: Terremoto e Società* (Garzanti).

The gradual emptying of the prefabs after the peak of the building program in May 1977 with 65,000 persons living there (Table III.1) showed different regional trends and pointed to social and demographic developments that might be typical for the aftermath of many more disasters.

2. Demolition

By 1990 all provisional dwellings in Friuli are supposed to have been demolished. Demolition costs of 30,000 lire per square meter will amount to $83 million more to be paid by the Italian government and the Region of Friuli-Venezia Giulia. Besides this expenditure, the problem of the subsequent use of both the fully developed settlement land and the dismantled prefabs arose.

Some prefabs were sold to private owners, some dedicated to public use (sport clubs and the like), and some might also have been exported to other disaster areas, such as Irpinia in the hinterland of Naples.

The area of the biggest baraccopolis of the region, Osoppo, is supposed to be used for a memorial park, because it is public property. Other communes will return the confiscated land to its former owners for agricultural purposes. The supply of fertile soil is a special problem. Publicly owned areas of former baraccopolises are to be earmarked for residential land use, but in the face of an oversupply of housing the question remains as to who should build there.

Table III.2 Persons in prefabs within the research area (51 communities)

	5.77	3.78	10.81	10.85	6.86	2.88
Persons in prefabs	58,091	56,934	34,163	23,925	16,907	5,200
Percent of total population	45.4	44.5	26.7	18.7	13.2	4.0

Source: Geipel, 1977, S.G.S, 1985, 1986; ISTAT, 1984; Regione F.-J.V. 1979-1985, Corriere della Sera 2.2.88, p. 7.

Photo 1: Predominantly abandoned prefabs settlement in Cavazzo Carnico, with remains of a donated South Tyrolean type. New buildings in the background.

Photo 2: Different prefab types juxtaposed: wooden houses from a company in Fellbach near Stuttgart, called "Stoccardo" by the Friulians, and Canadian containers of the Atco type.

3. The Problem of the "Squatters"

In the 51 communities surveyed with 127,000 inhabitants in 1977, more than 58,000 people had to live in prefabs until their homes were repaired or newly built. As soon as they made the latter move, from hut to home, and once they had received the last installment of their grant-in-aid, the victims were no longer entitled to stay in a prefab for which they had to pay neither rent, utility, maintenance, nor repair costs. These costs were covered by the Emergency Commissioner and/or the Friuli–Venezia Giulia Region. Many families, however, continued to take advantage of the possibility of living in a prefab, even when their new houses were ready for occupancy (Table III.2).

Some of the prefab types were of such good quality that they were fit for long–time use. In part they were upgraded for comfort, amenities, and building standard by private initiative (see Geipel, 1982, p. 118 for the types of prefabricated houses).

Hence, many huts continued to be used even if the family already owned its new house, even if merely as a second home or a residence for grandparents. This separated the several generations of the traditional Friulian family, allowing parents and young people to stand on their own feet. Thus, the modernization process in family life took advantage of the prefabs as a vehicle for emancipation.

On the other hand, empty but still habitable prefabs attracted people looking for a place to live, even if they had not been victims of the earthquake. A general lack of housing made prefabs attractive as a free-of-charge abode, especially for those who were too young to qualify for compensation money. Therefore, the prefab dwellers of the later period have to be divided into two groups:

1. *Occupants with title deed ("con titolo"):*

> a) people who lost their homes, acquired a prefab, and still live there because they still lack a reconstructed or repaired final place to live.

2. *Occupants without title deed ("senza titolo"):*

> a) persons who already had used up all of their compensation, and therefore should have moved to their new home
>
> b) persons untouched by the earthquake and who moved into an empty prefab only later.

Prefabs used by these squatters must of course be considered occupied an cannot be removed.

The duration of prefab occupancy in Friuli is therefore related not simply to the general efficiency of reconstruction, but to a considerable extent also to the way prefabs are being used. Evidently the utilization of emergency housing extends beyond the time necessary to house homeless victims of a disaster.

The Extent of the Squatter Phenomenon

Table III.3 shows the extent to which the squatter phenomenon could be observed in all communities of Friuli.

Table III.3 Families in prefabs: Total, and those with legal title

	1978	1980	1985	1986
Families	26,188	17,448	10,456	8,206
With legal title	26,188	10,950	3,955	2,901
Percent	100%	62.8%	37.8%	35.4%

Source: S.G.S., 1986, 1985 (a), 1983.

At least a good third of the prefab occupants as early as 1980 lived in their huts without title deed, and by 1986 this position grew to include two thirds of all families. The evacuation of the baraccopolises would have taken place much more quickly except for those who lived there without title deed. The proportion of such persons varies among the four categories of communities mentioned before (see Fig. III.2).

This share was consistently high in active communities with large populations at any time, but always less than in the passive communities. The proportion of the squatters grew steadily as time went on, but in economically strong communities more markedly than in the weak ones (Fig. III.3).

This means that the squatter problem was greatest in the most attractive communities of Friuli. More than two thirds of the prefab dwellers were living there without title deed in 1985, mostly in the industrialized communities of Osoppo, Maiano in the south, and Tolmezzo in the north.

The more inhabitants a community had and the better its location (foothills), the more quickly the prefabs have been emptied of their entitled occupants, but the sooner the squatters have been moving in. Eight years after the occupancy of emergency housing, the spatial pattern of the squatter phenomenon mirrors exactly the socioeconomic differentiation of the former disaster area. The North–South gradient clearly shows up. Figure III.3 shows the differences in socioeconomic development, with the active hill communities having a bigger share of squatters. Adjoining communities with good communication lines to these work centers also show this neighborhood effect by a higher share of squatters (Trasaghis, Forgaria, Montenars, Lusevera; Villa Santina close to Tolmezzo).

Some communities such as Dogna and Chiusaforte have their high share of squatters because of the building of the freeway between Austria and Trieste. The prefabs were used as accommodations for the construction gangs. This is

Fig. III.1

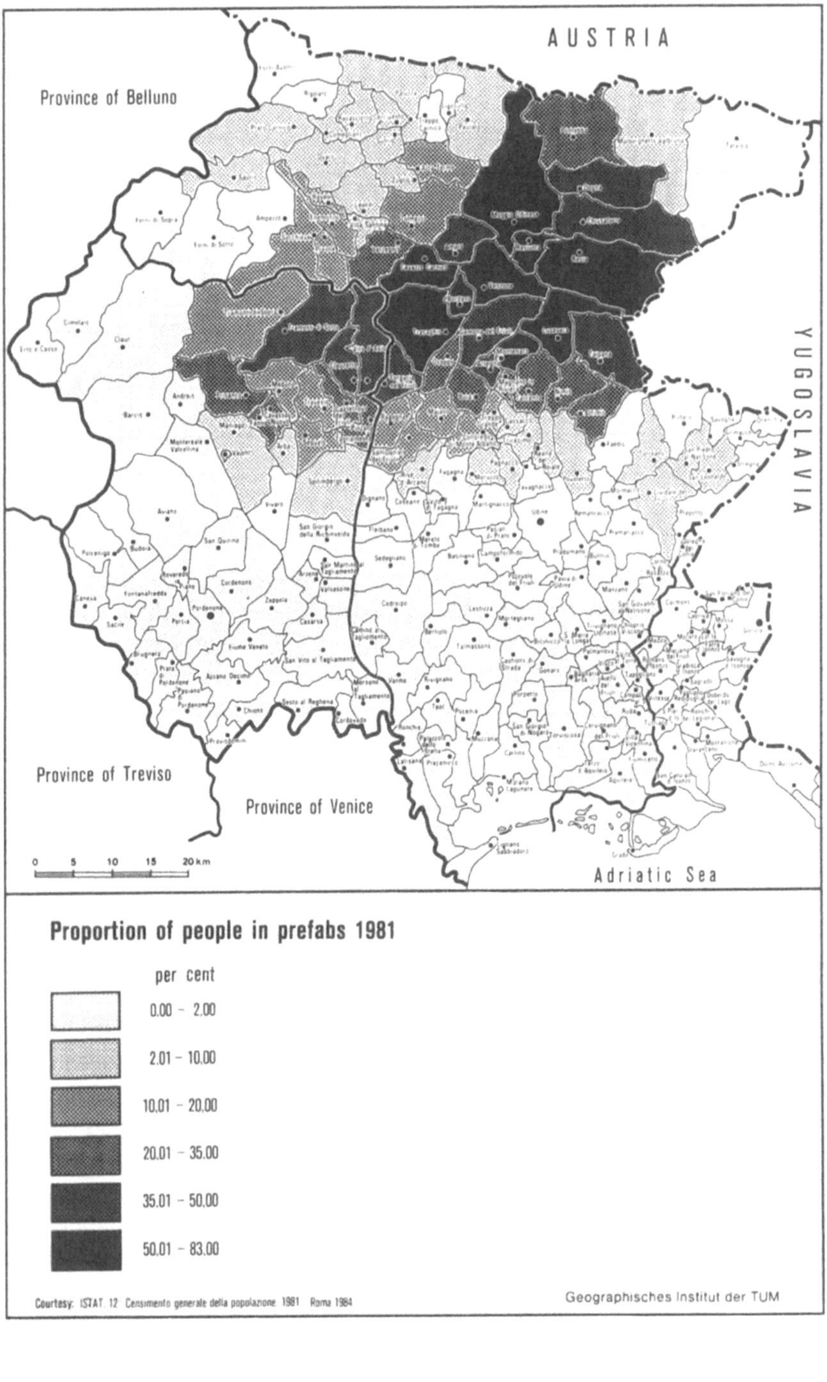
AUSTRIA
Province of Belluno
YUGOSLAVIA
Province of Treviso
Province of Venice
Adriatic Sea
0 5 10 15 20 km
Proportion of people in prefabs 1981
per cent
0.00 - 2.00
2.01 - 10.00
10.01 - 20.00
20.01 - 35.00
35.01 - 50.00
50.01 - 83.00
Courtesy: ISTAT. 12 Censimento generale della popolazione 1981 Roma 1984
Geographisches Institut der TUM

similar to the squatters in Val Belice, now that reconstruction finally has taken place even there. They are mostly North African migrant workers. Their predicament, and local attitudes to them are being investigated at the moment by researchers from the City University of New York (communication,. Alexander).

Reasons for the Squatter Occupancy

There are two main reasons for the squatter occupancy:

- The high quality of many of the prefabs in Friuli rendered a long-time occupancy possible. The huts were acceptable for dwelling. (This suggests a lesson for the future: Don't build too well!) Minister Zamberletti (1986) and chief planner Chiavola (1985) justified this good quality with the argument that the occupation of the prefabs for an expected period of several years would require some measure of comfort.

- The world–wide recession of the 1980s affected a large part of the population in Friuli, too, quite apart from the earthquakes. The growing number of people on welfare led to a high demand for cheap housing.

On the other hand, after the earthquake it was mainly industry and private building that flourished in Friuli. Jobless elder persons without title deed were at a disadvantage. For them the ample supply of empty prefabs offered an acceptable source of housing rent-free because the communities preferred to solve their social problems by directing welfare applicants into empty prefabs. Especially in parts of Friuli where jobs were available, which in any case were target areas of intraregional mobility from mountain to foothills and plains, the prefabs were used in this way. The lack of publicly sponsored housing programs could be "solved" in this way thanks to the earthquake and its aftermath. House hunters could be pacified during a transition stage, pushing the real problem solving into the future.

Pretty much the same is the case with elderly people who lost their supporting relatives. As early as 1977 a social plan for old people was called for (Geipel, 1977, p. 182) in order to avoid their ghettoization in the baraccopolises. But that was exactly what happened, at least to some extent. The social development of prefab occupancey is an indicator of the fact that, after a disaster, the previous social structures persist. Adjustments to the disaster soon lose their original function and become part of a general social development independent of the catastrophe. Care was not taken to make sure that once

Fig. III.2
Percentage of families living in prefabs possessing legal title, by type of commune

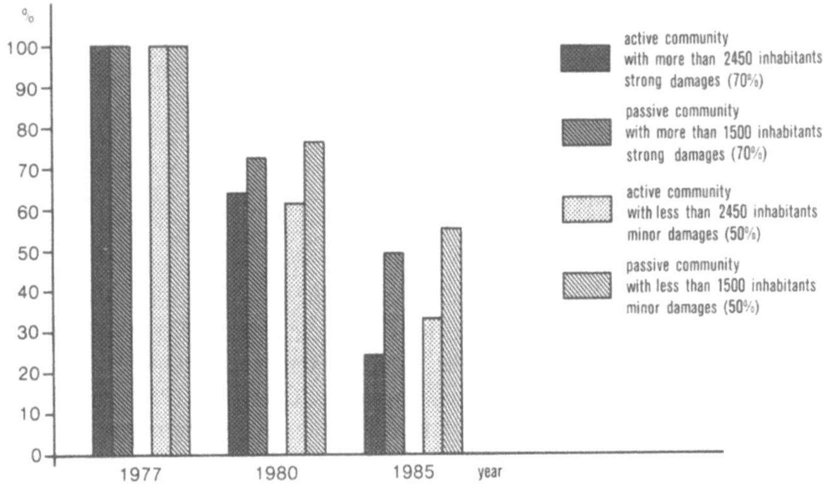

Source: own results

Fig. III.3
Percentage of families without legal title, according to differences in socio - economic status of commune

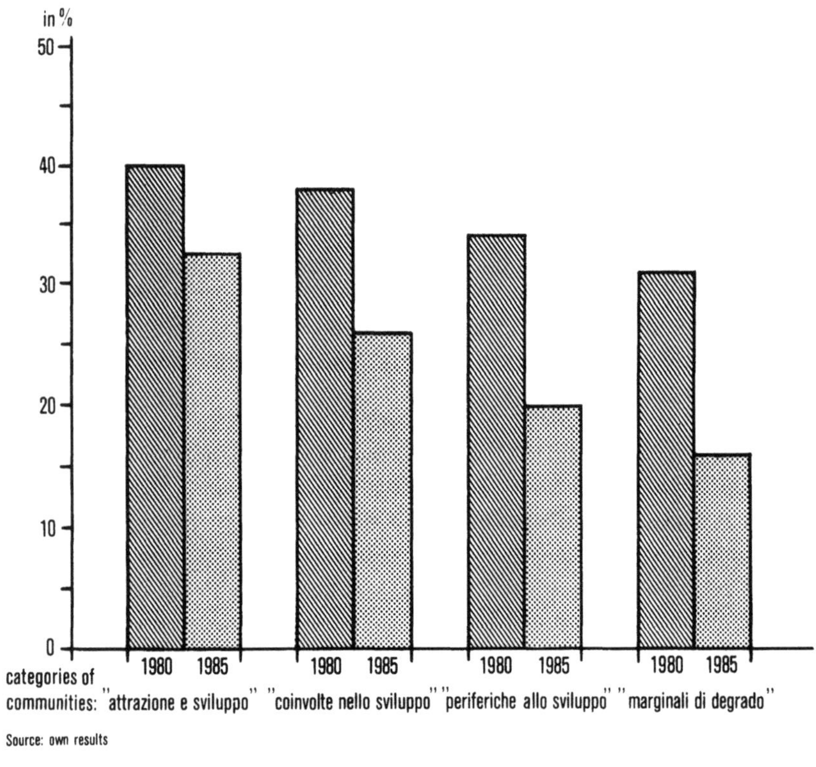

Source: own results

ready-built homes were provided, the occupancy of prefabs would be terminated. More efficient supervision or other suitable measures should have been enacted in order to make leaving the huts more attractive. Already, with the construction of emergency housing, the policy of abolishing them afterward should have been maintained and enacted by law.

4. The Composition of the Population in the Prefab Towns

The findings of hazard research show that "elderly victims are more likely to experience a long term drop in standard of living than are young victims" (Bolin, 1982, p. 236). ". . .those over 63 years of age suffered a proportionally greater level of fatalities" (Cochrane, 1975 p. 37).

Changes in age groups in the prefab towns in comparison to the total population of a community could be studied in only five communities because of high research costs. They had all been heavily destroyed, but differed in size and socioeconomic development. We distinguished between two groups (type A = good structure, as exemplified in Osoppo and Artegna, and type B = weakly structured with the examples of Montenars, Resiutta, and Bordano). We had to distinguish also between two periods:

before 1980:

- in the active communities (type A) the working population (25–54 years of age) had evidently left the prefabs, being underrepresented in comparison to the whole population. The older age groups (older than 54 years) were clearly overrepresented in the huts.

- in the passive communities (type B), in contrast the age structure was the same in the baraccopolises and in the communities as a whole. No demographic selection took place.

1980 - 1985:

- in both communities of type A, the prefab dwellers included more old people (70 years and older) and people just under the pension age (50–65 years), those 30–45 years of age in turn were underrepresented.

- in the three communities of type B, only people over 74 years were grossly overrepresented in the prefabs. All other age groups showed no differences.

In the prefabs of Friuli the elderly are evidently a social problem group. Their bad financial situation, lack of initiative, and generally isolated social position prevented them from participation in a fast and successful reconstruction of their homes (Stagl, 1980, p. 60).

As a consequence, they form the group of prefab dwellers who stayed longest in their huts. Therefore, the appearance of the late baraccopolises was marked by the presence of old people, especially old women. Again, at the end of the prefab cycle, what became obvious at the beginning of the disaster itself was repeated: the greatest sacrifice falls on the older population in natural disasters. This is clearly shown in the monument to the victims in the church Sta. Maria delle Grazie in Udine, which indicates sex and age of the dead (Fig. III.4).

The divergences between Friuli's normal age pyramid and that of the victims' reflect partly the young soldiers who were killed in the barracks of Gemona, but moreso the losses in the age group, who older than 60 years, mostly lived in the older buildings in the old city cores. In the baraccopolises, too, it is they who form the rear guard on the way back to normality.

Table III.4: Age structure at various periods, by type of community

Type of community	1980	1985
Type A High destruction High poulation Good socioeconomic development	Overrepresentation of old people Underrepresentation of active age-groups (25-54 years)	Overrepresentation of old people Underrepresentation of active age groups (25-54 years) Growth of age group 50-60 Overrepresentation of young people (20-25 years)
Type B High destruction Low population Bad socioeconomic development	Age structure in prefabs correlates with that of the whole community	Distinct overrepresentation of old people Other groups like average of community

Fig. III.4
Age and sex characteristics of the dead from the May 1976 earthquake

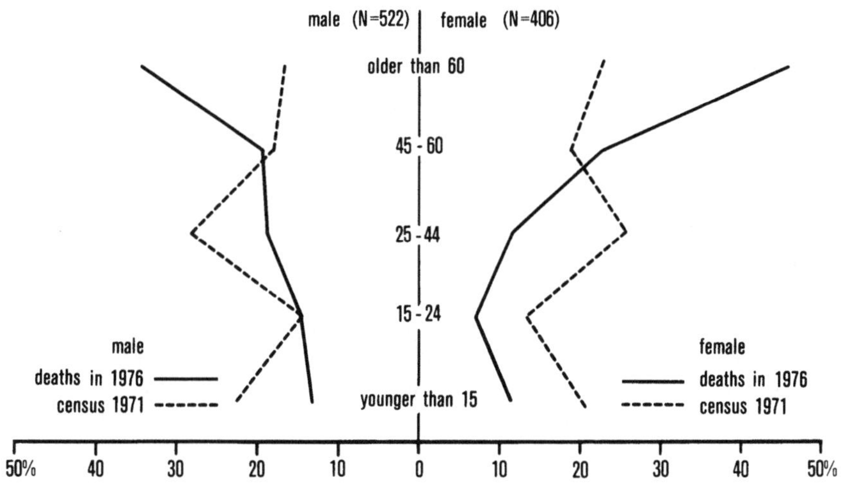

Sources: Data from thememorial in the church, census 1971,ISTAT 1973

5. The Social Selection Process in the Prefab Towns

Old people are always a particular problem group in the prefab towns, and among them especially those who live alone. They stay longest in the huts, and possess title deeds. They form the "leftovers" in the huts, who even 10 years after the quake could not use the compensation money due to them in a way that enabled them to leave the huts. For many of them these will be their home forever. This could be observed in all prefab towns of Friuli. The second problem group among the prefab dwellers were of another kind. In the economically attractive communities of Friuli a population of squatters had formed, consisting of

- young families, using the possibilities offered by empty huts to form their own household

- families whose house was ready but who for various reasons preferred to stay in the prefabs.

Ten years after their construction, the emergency housing facilities were occupied by two very different groups each for different reasons:

- one group, which fell through the holes in the social network, and belonged to a marginal population without hope of overcoming the emergency situation of the prefab.

> one group at the beginning of their working careers, who have often established young families, for whom staying in a hut means a mere transition period until an apartment can be rented somewhere and "normal life" started.

Some of the young people are immigrants. Even in Friuli, with its surplus of habitable space, there is a shortage of cheap housing, especially in the work centers. The prefab towns showed a natural degradation because of overuse. Social structure as well as appearance indicate that baraccopolises are slums of hope and despair.

Those living here without title deed for an interim time do so in the hope of moving soon. The others are full of resignation. In some cases the grandparents have left their compensation grants to the young people and (now without title) live here as they might have done in an addition to the farm house before. In Friuli as in so many other places, it will not be easy for single, old persons in need of aid to leave the huts for old-age homes.

The instrument of "Einliegerwohnung" (a small independent flat within a family home, built for the use of a single person) is not known in Friuli. But to be integrated into a family network is important for seniors. The productive age groups with higher income have been much better able to overcome "prefab life." There were no measures to prevent the reoccupation of empty huts or to make it unfeasible for families to defend such a preempted prefab despite having also just built a new home. This prolonged the use of what were intended to be only temporary facilities, and added to the slum character of the settlements. When we asked 6500 persons in 1977 what would become of Friuli, they knew who would have to wait for their death in prefabs: the old people. The goal of social restructuring that the experts had set could by no means be achieved, despite the otherwise positive experience with the prefab towns of Friuli. One example may illustrate what was said before: "Rauna II."

This small settlement, consisting of 15 prefabs with 44 apartments of the Della Valentina type, is part of the Oseacco faction in the community of Resia, an over–aged mountain settlement. One fourth of all households in 1977 consisted of single persons older than 60 years, and in 1980 and 1985 one-person households dominated in the prefabs. In 1977, the age groups of the 85 inhabitants were in accordance with those of the total community. In 1980, however, persons aged 60 years and older predominated, because the active age groups had left. There is no squatter problem because the community has no workplaces to offer. In the course of time the remaining inhabitants space themselves out more as prefabs are left empty, by others moving out to their new homes (Fig. III.5).

In 1985 there were only two real next-door neighbors left. Everyone who has come in later tries to keep some distance from the nearest neighbors.

Prefab dwellers, who complained about lacking privacy, have created it themselves whenever possible (Catarinussi and Pelanda, 1981).

6. Prefab Settlements as Indicators of the Phases of Reconstruction

Studies of the long–term effects of emergency settlements aim at more than the mere documentation of expected time span of use and of a specific evacuation process. It becomes evident that reconstruction following disaster in general, and especially evacuation of emergency housing, is strongly related to socioeconomic development and the spatial distribution of a disaster-affected society. After destructive disasters such as the two earthquakes of 1976 in Friuli, a reconstruction plan does not begin with a "tabula rasa" but has to take notice of persistent spatial structures. Kates et al. (1977) argue that the persistence of previous spatial structures is strong and that ". . .the time needed for reconstruction is a reflection of predisaster urban plans, the damage suffered, and the resources available for recovery. . ." (Kates et al., 1977, p. 262). Applied to the prefab towns in Friuli, this means that their evacuation in favor of permanent homes was influenced by certain specific variables and the regional differentiation thereof. On the other hand, the functional change of the prefabs, from housing victims to housing squatters, may serve as an indicator for certain phases of reconstruction.

In 1980 Hogg tried to apply the Haas–Kates–Bowden model (see below) to Venzone (Hogg, 1980, p. 178), and found it sadly lacking, maybe because of the exceptional status of the "National Monument," and because his period of observation was too short. We could extend this attempt for eight more years, observe the whole reconstruction process, and test it against the four phases of Haas et al. (1977).

The move into prefabs can be taken as the beginning of the phase "Reconstruction I," and the marked increase in the number of squatters in the huts as the beginning of the phase "Reconstruction II." This means that, five years after the disaster, reconstruction has proceeded so far that the emergency build ings are increasingly used by squatters. This may be interpreted as a further step toward the return to the predisaster level. In 1990 all prefabs will be demolished and all confiscated areas returned to their former owners (communication from Dott. Macchin, in charge of the Emergency Commissioner's statistical bureau [1989]).

May 1987 was the deadline for approximately 10,000 persons to move out of the prefabs (Messaggero Veneto, May 9, 1987). Half of them (5000) had already received their compensation grants, 3000 had forfeited their claims to

Fig. III.5

The example of "Rauna II" at three different stages. Occupied accommodations, with number of persons Accommodations where someone moved in or out. Accommodations occupied without legal title

The baraccopolis of "Rauna II" in 1977

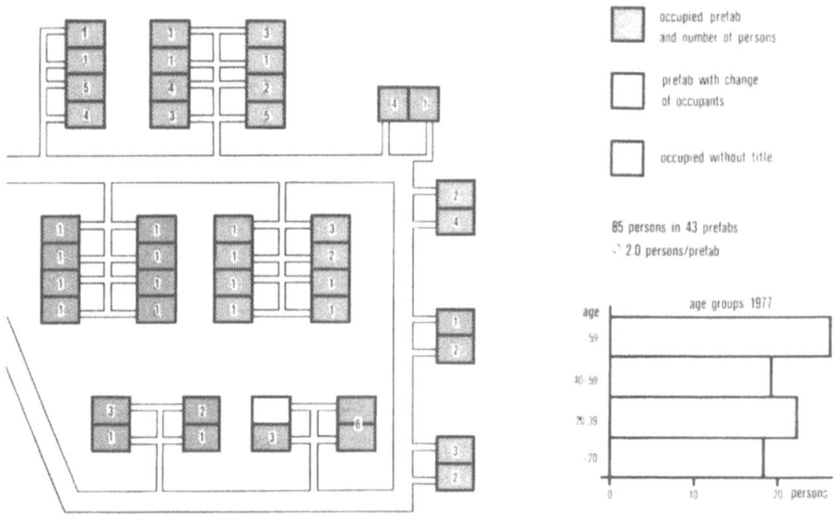

The baraccopolis of "Rauna II" in 1980

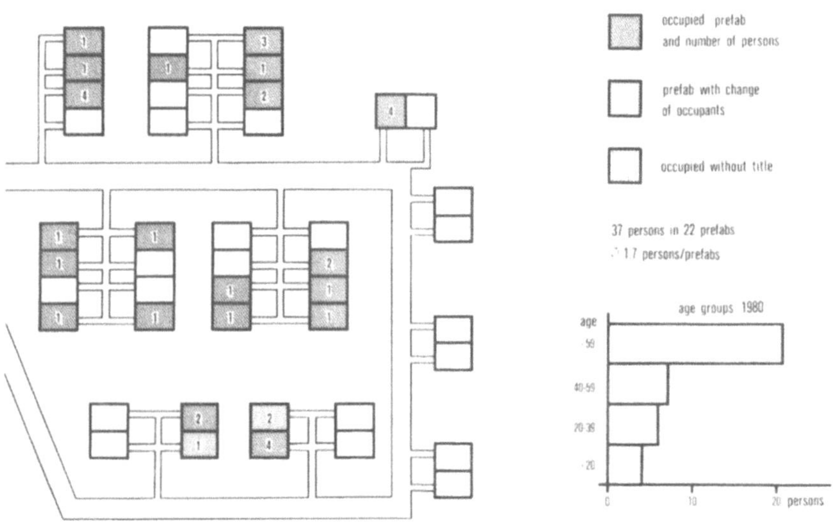

The baraccopolis of "Rauna II" in 1985

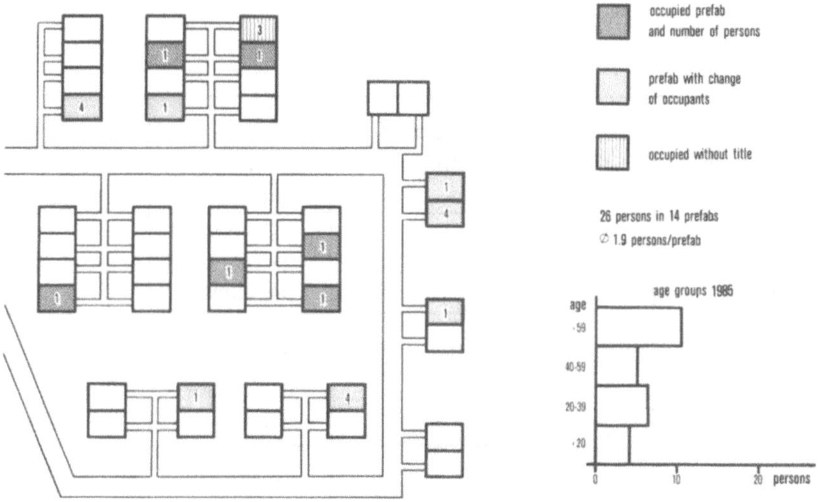

compensation money, and 2000 were cases for social welfare who had to be redirected to other institutions. Demolition of the remaining prefabs was accomplished in two steps (as tested in some communities):

1. concentration of all prefabs from all fractions at one point in the community,

2. final dismantling of this (only and last) prefab settlement in the community, lasting until 1990.

Table III.5 attempts to show some relations that were revealed during the evaluation of prefab statistics. Even if transferability of these findings to other disasters may not be possible in all cases, they should make us more sensitive to social problem groups, to the effect of spatial persistence, and to the expected duration of use.

Table III.5 Overview of the emergency housing situation

Initial situation	Socio-economic structure	Extent of damage	Recons-truction philosophy	Site and quality of prefabs
Use of the prefabs	Duration	Intensity	Structure of user population	-
Affected persons	Elderly	Young families	Arrivals from else-where	Social debility
Problems	Continuing costs for prefabs	Segregation	Distorted settlement structure	Excess of living space
Potential solutions	Building of public housing and old age housing	Communal and regional development planning	Controlled removal of prefabs	-

Fig. III.6 Comprehensive model of Reconstruction Phases in Friuli

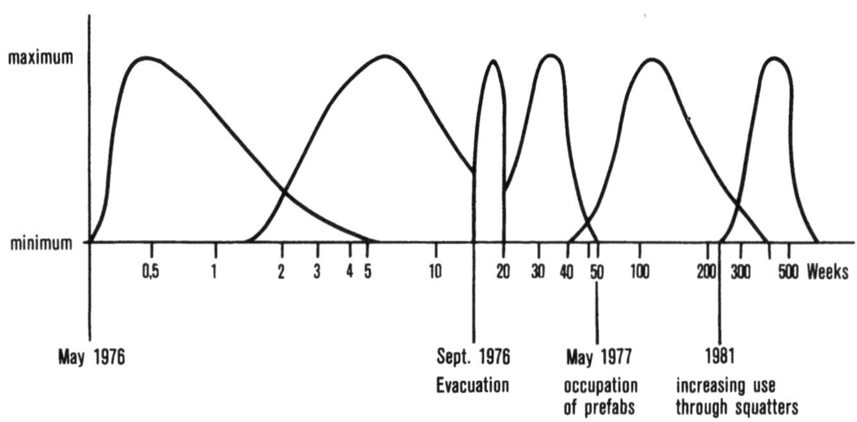

IV
The Single Stages of Reconstruction

Relations within the man–environment–system change with time after a disaster. Hazard research proposes that this time can be divided into four typical stages. Our research aimed to test this theory.

These four stages have been distinguished more or less clearly by Kates and Pijawka (1977) in their central chapter "From Rubble to Monument" in *Reconstruction Following Disaster*, a still very exciting book. This four-stage model should be quoted at the beginning of our fourth chapter (Fig. IV.1) because we want to test it against the reality we found in Friuli between 1976 and 1988.

Kates and Pijawka postulate ". . .that each of the first three periods lasts approximately 10 times longer than the previous one" (p. 3). Altogether, the examples examined took approximately 500 weeks (10 years) for the total rehabilitation of all vital relations.

These stages were derived deductively from the examples of San Francisco (1906 earthquake) and Alaska 1964, where reconstruction has been finished

Fig. IV.1

Model of Reconstruction Phases in Friuli

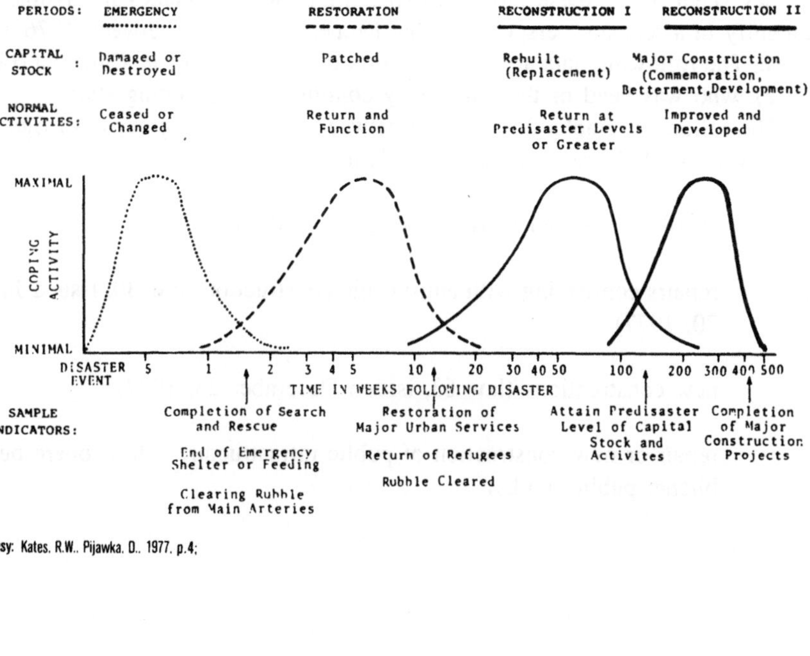

Courtesy: Kates. R.W.. Pijawka. D.. 1977. p.4;

long since, and two projections into the future: the Rapid City Flood of 1972 and the Managua earthquake of the same year. But there was a lack of indicators that would allow us to limit the beginning or the end of the singular stages unequivocally.

For reconstruction planning, it is important to recognize differently structured stages, their duration, and specific problems. This not only facilitates a more efficient assignment of funds, but it can also serve as an indicator of possible difficulties.

1. Reconstruction Stages and Expenditure of Funds

To our knowledge Friuli is the first greater disaster in history where the four stages can be tested empirically. In Friuli 70% to 80% of all funds dedicated to reconstruction were pooled by the emergency commissioner ("Segreteria straordinaria") with a precise accounting day by day, village by village, and person by person: Rumors of fraud in previous disasters made this bookkeeping a challenge for the authorities in charge of the Friuli disaster. Therefore, the spending of funds exactly mirrors the reconstruction, because everybody who received compensation had to prove his title by his progress in restoration work. A building finished in the rough could claim 50% of the compensation money, the ready building 100%. If we define the four stages in order to measure them, stage one means emergency measures, stage two a functioning society (although under difficult conditions), stage three the physical reconstruction of the vital spheres of housing and workplaces, and stage four the complementary construction activities from paving to church building. The money spent on all these different expenditures in time makes it possible to quantify in a realistic perspective what happened in Friuli between 1976 and 1988. The following data are a contribution of our co-author E. Chiavola (Fig. IV.2), who was head of the emergency commission's planning staff.

The following graphs differentiate between five budget groups; all figures are given in 1977 lire to adjust for inflation:

1. small repairs under law 17, of May 7, 1976

2. repairs confirming with antiseismic construction (law 30, issued June 20, 1977)

3. new constructions (law 63, issued December 23, 1977)

4. repair or new construction of public infrastructure (OP = opere pubbliche: public works)

5. repair or new construction carried out directly by the emergency commissioner.

Law 17 also prescribes expenditures for the establishment of the prefabs, hereby including the stage of restoration. The curve for these expenditures rises steeply but also falls in the same steep way. This stage was essentially finished in May 1977, seven and a half months after the second earthquake in September 1976, when the evacuees had to return from the seaside resorts to clear the hotels for summer tourism, one of the main sources of income of the whole region.

At the end of 1977 and with the beginning of 1978 the two main processes of private reconstruction began, under the provisions of law 30 for repairs and law 63 for new building. Part of the Reconstruction I stage, their expenditure curve also rises steeply and reaches its peak at the end of 1979 and the beginning of 1980. Compared with law 17 (small repairs and prefabs), the curves are much smoother.

During 1980 the regional administration began rebuilding public infrastructure (e.g., schools moving out of emergency buildings, construction of municipal buildings and town halls). This curve rises slowly, so that its maximum is not reached before the end of 1983. Somewhat earlier, at the end of 1979, other repairs and new constructions for infrastructures (PW) began. These two curves can be associated with stage Reconstruction II.

Indirectly, stage I (emergency) is also included in these curves. It starts with the disaster and ends with the erecting of huts, where the accounting of the

Fig. IV.2 The Five Reconstruction Phases in Friuli

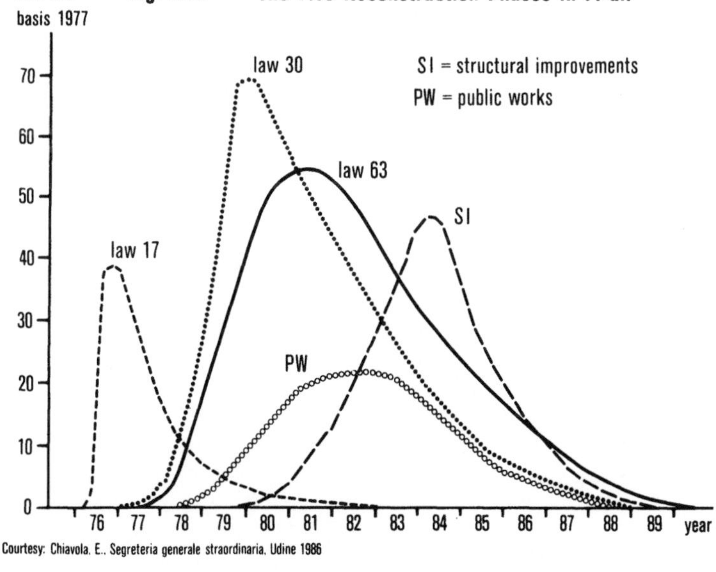

Courtesy: Chiavola, E., Segreteria generale straordinaria, Udine 1986

Fig. IV.3
Total Expenditures of the Regional Government for Reconstruction in bill. Lire

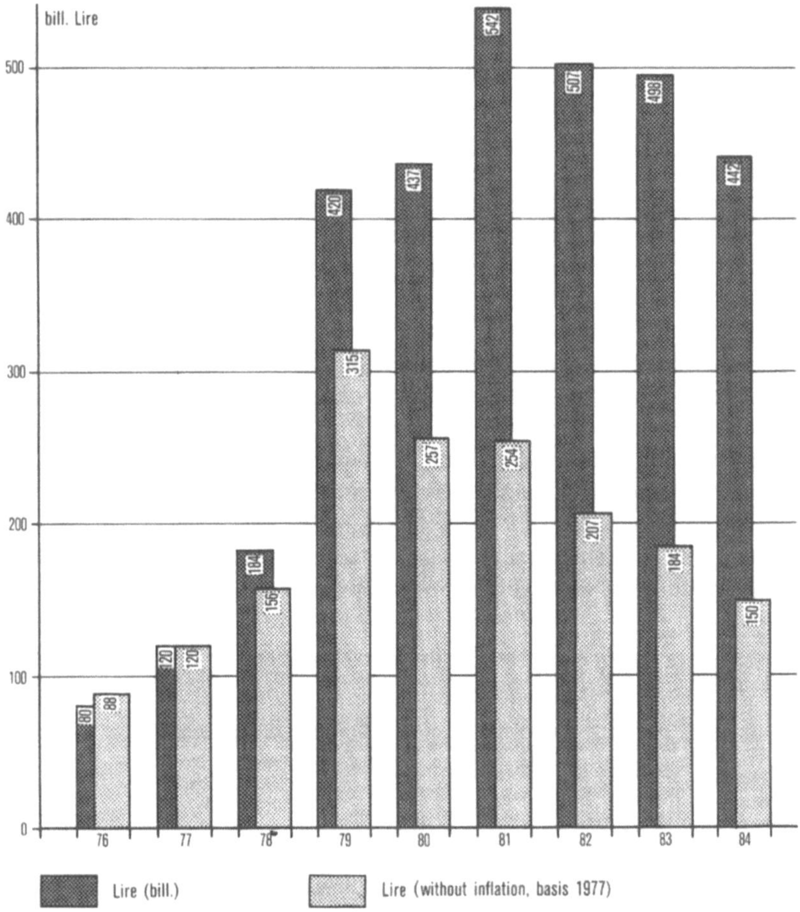

Source: Regione Autonoma Friuli - Venezia Giulia. Relazione sullo stato delle attività regionali per la ricostruzione delle zone colpite dai sismi del 1976. Trieste 1985. p. 23

emergency commissioner started. Figure IV.2 shows that the single types of expenditures related to different activities roughly correspond to the theoretically postulated stages. If we consider only the total of reconstruction expenditures (Fig. IV.3) we find merely a curve of a normal distribution, indicating that every year up to the 1979 normal peak more money was spent until the volume of expenditures decreased little by little. Breaking down this curve to reflect the funds raised by the various laws we gain insight into the sequence of stages.

2. Critical Moments During Reconstruction

Combining the expenditures of all five categories, we find that reconstruction follows a typical rhythm. The first crest of the wave is the stage of erecting solid emergency buildings. A wave trough follows, before the expenditures for the building of permanent houses and public amenities climb to a similar crest, lasting longer than the previous one (Fig. IV.4).

The curve of expenditures during the stages after a disaster, when all the altruistic gifts and donations have been drained and international solidarity ceased, show the real activities in a disaster area (Fig. IV.5).

Fig. IV.4

Total Expenditures of the Regional Government and of the Emergency Commission

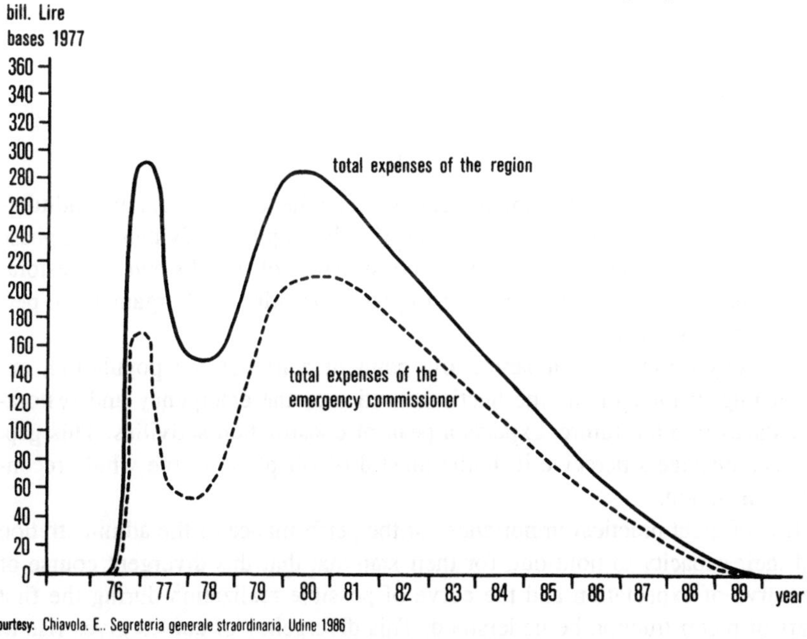

Courtesy: Chiavola. E.. Segreteria generale straordinaria. Udine 1986

Fig. IV.5
Difference between intended and actual volume of expenditure

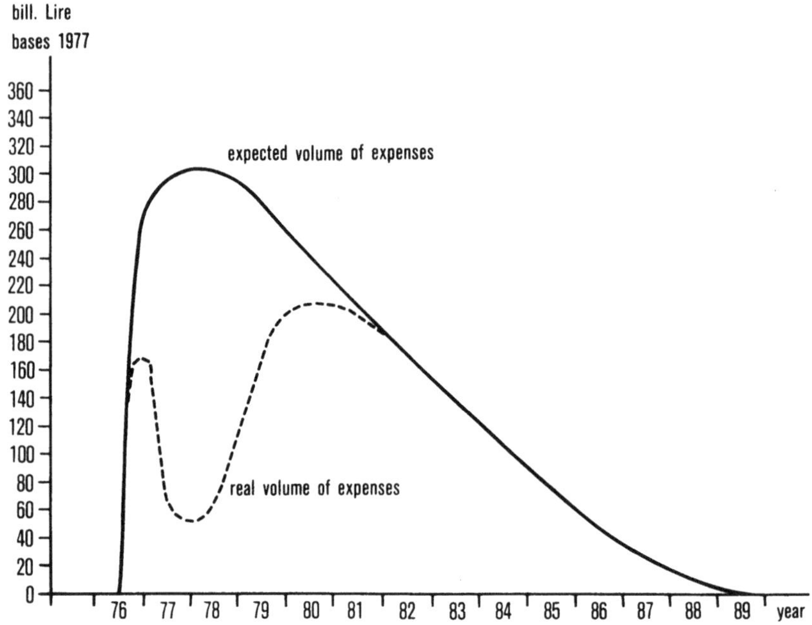

Courtesy: Chiavola. E.. Segreteria generale straordinaria. Udine 1986

Notable in Figure IV.5 is the big gap between the levels of real expenditures and the expectations of the victims and of public opinion. Evidently, a phase of consolidation is necessary after the construction of prefabs and the restoration of major urban services in order to plan, determine, and organize the permanent reconstruction.

Precisely during this apparent low point of activities the population, extrapolating in their minds the hectic activities of the emergency and restoration stages into the future, expects a peak of construction activities. This gap, as crisis managers perceive it, is the most difficult phase of the whole reconstruction period.

It is of great practical importance for the performance of the administration and their capacity to hold out, for their stamina, that this divergent course of the curve of expectation and the curve of possible realization during the first years of reconstruction be understood. This divergency of curves gives rise to four dangerous situations, and the manner in which they are dealt with is decisive for success or failure of reconstruction.

Fig. IV.6
The Four Critical Moments during the Reconstruction Process

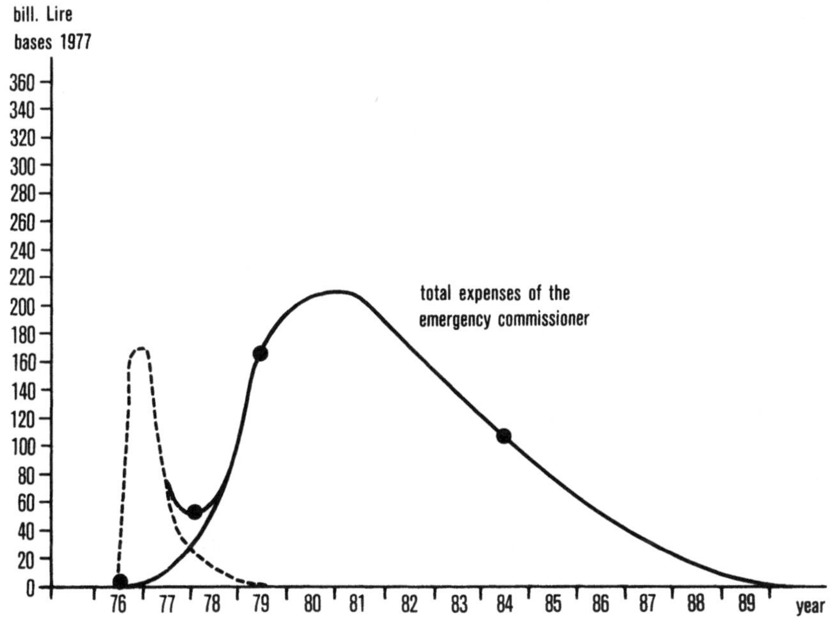

Courtesy: Chiavola, E., Segreteria generale straordinaria, Udine 1986

Again we can consult the diagram, showing expenditures in the course of time (Fig. IV.6) and depicting what is really happening in the visible rebuilding of towns and villages. It reveals the four critical situations.

The first critical point, of course, is the disaster itself, where effective emergency measures have to be taken to carry the victims beyond the shock of the events and convey a sense of optimism for the future. In this respect Friuli, because of the "five lucky constellations," was in a comparatively good situation.

The second critical phase (with a duration of approximately 18 months between the middle of May 1977 and the end of 1978) lies exactly between the summit of the restoration and that of Reconstruction I, two stages that independently follow each other. The minimum between the neighboring maxima represents the boundary line between two processes and in Friuli was concomitant with calling back into office the previously dismissed emergency commissioner in Spring 1977. After the dynamism of the first half year, the victims insisted on accomplishing the reconstruction, building houses, and on generally greater activity. At this moment the necessary laws had not yet been issued, modalities of the payment of compensation not yet settled, the necessary controls not yet established, and plans not yet worked out. (All those measures of course could have been prepared before a disaster as a precaution, but this is seldom the case in possible disaster areas, as the collapse of highways in the San Francisco area shows.) The forced idleness of victims, capable and ready to act on their own, leads to accusations against administration, illegal activities (illicit buildings), and apathy. In many disaster situations, reconstruction never passes this second point. It stops as soon as the international aid–giving organizations have left the scene. Reconstruction must be considered as having failed. Only through a strict control by an able administration, capable of enduring stress and of great stamina, can this second point of danger be conquered.

The third critical point occurs when reconstruction is running full speed and the opinion prevails among the victims that this will go on and on. This problem has a technical and psychological cause. Excessive demand, compared with insufficient building capacity, leads to a cost explosion and corresponding difficulties in private and public construction. Numerous building firms were founded overnight to reap fast profits. This was facilitated by the fact that many Friulians worked as masons in construction gangs all over Europe and overseas. Conferring of contracts became chaotic. In 1979, with a general inflation rate of 21% construction prices mounted to 30% and more above valid contracts, some exceeding 80%. The share allotted to wages under law 30 (repairs) in 1978 amounted to 72% (Messagero Veneto, Nov. 11, 1978). Figure IV.7 shows the excess of building costs above inflation between 1980 and 1982.

In the second half of 1980, 18 major construction firms from outside the region were hired by the authorities to fulfill collective contracts (ca. 15% of the whole construction volume) in order to break up the local cartels; this gave rise to angry protests against the "alienation from abroad." In Friuli, a "building orgy" had prevailed, described by secretary general Chiavola as follows: "The inhabitants, desperately and almost morbidly clinging to their houses, used up all their resources including normal credits, to repair, to enlarge and to embellish" far beyond what had been authorized (Chiavola, 1985). The secretary general thereby destroys the myth of the small, brave mason, who, out of sheer love of his craft and dedication to his home village, despite low wages repairs one house after the other carefully, while using materials and antiseismic techniques hitherto unknown to him.

The 2500 builders from abroad stabilized the market. The domestic firms felt "ruined," but building prices returned to a level below the general rate of inflation. On April 30, 1985, this intervention was terminated, which brought the construction business in Friuli back to a seller's market with discounts of up to 20%. But it heightened the tensions against "Rome" and Roman centralism, because many architects from abroad designed buildings following a cosmopolitan style and through this deprived Friuli of some of its identity.

Finally, there is a fourth critical point, which is connected with the expi-

Fig. IV.7
Percentage increase in building costs, compared with Inflation

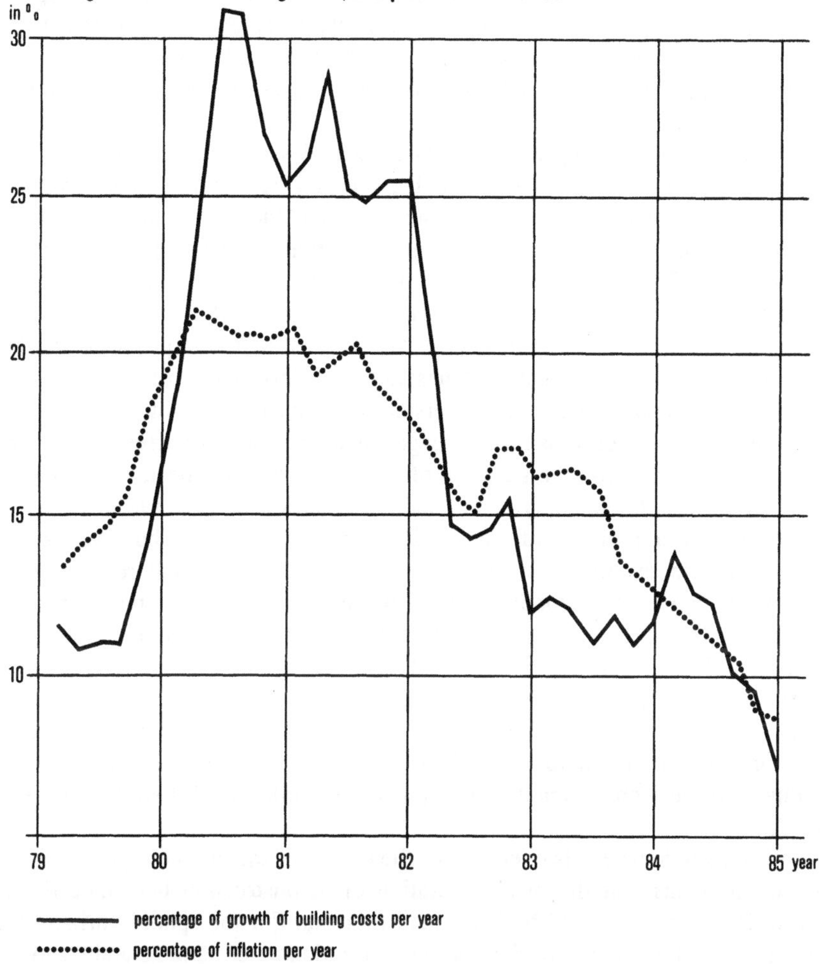

Courtesy: Regione Autonoma Friuli - Venezia Giulia. Relazione sullo stato delle attività regionali per la ricostruzione delle zone colpite dai sismi del 1976. Trieste 1985. p. 26

ration of the reconstruction process. Most of the problems associated with it stem from the cases (10% in Friuli) that for whatever reasons, have not yet found a solution. This concerned not the weakest groups of society alone (old and/or handicapped people who fell through the social safety net), but also those who took too high a financial risk and now asked for an additional compensation, or those (especially old persons) who did not get any compensation or forfeited their titles out of carelessness. Finally, firms are concerned that after the end of the building boom lost their clients, as well as workers threatened with unemployment.

The desperate and morbid "craving to build" in Friuli derived its special character from the fact that for seismic safety reasons, it surrendered to the "logic of concrete" (Hackelsberger, 1985). A "steel and concrete province" of a depressing, fortress-like bulwark character was erected to contend with the risks of life in this area.

"Concrete, on its progress from plastic shapelessness to the definite, predestined shape of casehardened solidity becomes the key material that depicts the general situation in a quite susceptible way. It is the irrevocable character of founding concrete which depicts everything that happens uninfluenced and irreversibly. The perception of this nature of concrete coincides with the ubiquity of this building material" (Hackelsberger, 1985).

Under the traumatic experience of its total destruction, a province that had handled building with quarry stones and boulders with excellent dexterity turned to concrete. "Whoever understands the psychology of concrete, who experienced its determinism, who engaged in the game of irreversible hardness, may comprehend the situation of all those who have been condemned to defencelessness by wrongly used concrete" (Hackelsberger, 1985).

Friuli, for a period, was an El Dorado for architects, but the balance achieved with the new construction is ambiguous. Historical reconstruction "stone for stone" (Venzone), modernistic display piece buildings, especially city halls (Osoppo), and the adaption of the new to the old (Gemona) are counterpoised. In the more remote heads of valleys, on the other hand, a much less luxurious functionalism, a do-it-yourself style of the master masons prevails. But the construction as a whole is now much less seismically susceptible: three tremors measuring up to Rm=4.1 on February 1, 1988 did not cause any damage to the reconstructed houses ("Corriere della Sera," February 2, 1988, p. 7).

We conclude here this overview of the course of the disaster cycle in Friuli, which confirms that the overall duration of reconstruction here indeed took those 10 years postulated by Kates and Pijawka. For the spatial variations of these stages, however, especially in respect to degree of damage and socioeconomic development, we proceed to the next section.

3. Regional Differentiation of Reconstruction

Following Rubin et al. (1985), differences in reconstruction are mainly derived from:

- the nature of the disaster agent

- the extent of the damages

- the availability of resources (human and material) (Rubin et al., 1985, p. 7).

As for the nature of the disaster agent during an earthquake, intensity and environment (physiogeographic conditions, age of structures, building material or the exact time of the event, e.g., rush hour) are of greater importance than the type of disaster itself.

In the case of Friuli, the nature of the disaster agent is everywhere the same, but the extent of the damages differs, as does the availability of resources. This is not true in respect to financial means in relation to damage (laws valid for the whole area) but in respect to the human and local preconditions. There are differences as well between as within the communities. In Osoppo, for instance, reconstruction in the outskirts started as early as 1977, but in the historical center (Centro Storico) only in February 1979. In Bordano, after planning that lasted from 1977 to 1979, the first repairs were carried out in November, 1979, and the first private activities did not start before 1980. Smaller, mostly remote mountain communities needed more time before they could start with rebuilding. Among the 45 destroyed communities, 13 had still not received any compensation for new construction under law 63/77 in 1978 (Fig. IV.8).

The main planning stage started in Fall 1978 and lasted until the beginning of 1981, when construction started. The completion of buildings was relatively evenly distributed with repairs going on (law 30/77) from the middle of 1980 to the end of 1983 (cf. stages of Artegna, Fig. IV.9).

Figure IV.9 also shows how strongly building around the year depended on weather conditions (with an intermission in winter) and other influences, for instance, the possiblity of writing off taxes for the whole year if work was completed by December. Cuny (1983) shows that the same was true in Guatemala (Fig. IV.10).

The average duration of building under closer survey lasted (independent of private, cooperative, or public construction) 25 months in four communities. In Artegna, where rebuilding was especially fully documented by the commissioner, the time between 50% payment (start) and 100% (finish) lasted 25.8 months. But predisaster situation, social status, and features of the locality (center or periphery) as well as building strategy (private or public) differentiated the program to a marked extent (Fig. IV.11).

Very often, in reports on disasters, destruction alone, and not socioeconomic predisaster development or size of community, are taken into account. How these variables influence reconstruction will be discussed in the following paragraph.

Fig. IV.8

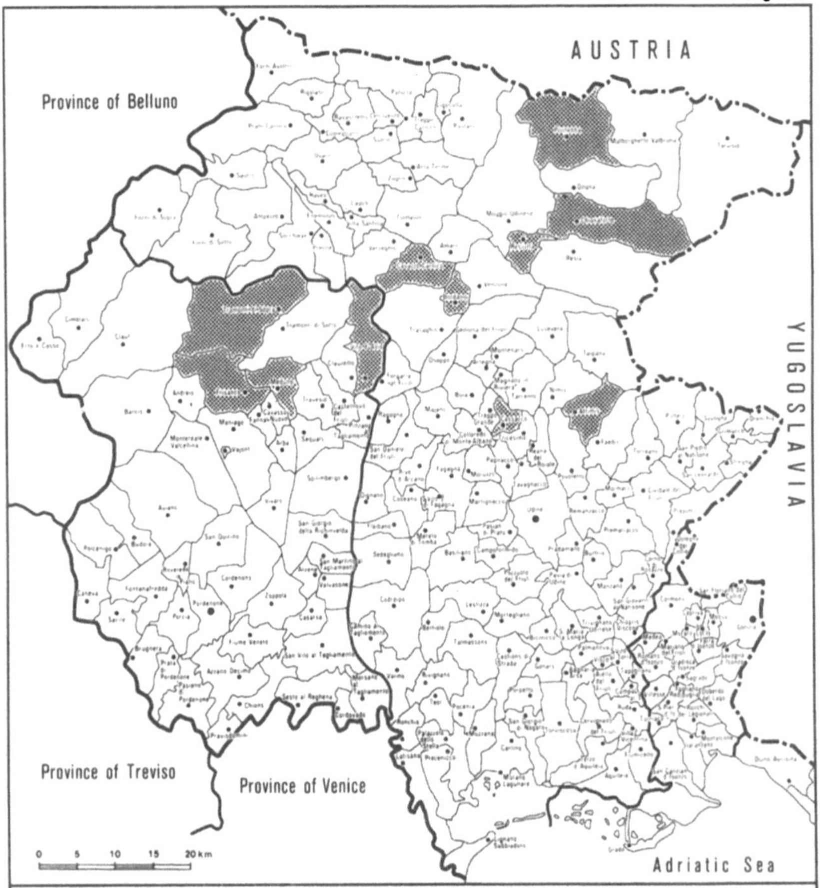

Communes in which new construction began only in 1978

Courtesy:
Regione Autonoma Friuli - Venezia Giulia. Relazione sullo stato delle attività regionali per la ricostruzione delle zone colpite dai sismi del 1976. Trieste 1985. Allegato B

Geographisches Institut der TU München

Fig. IV.9
The rhythm of new building and repair work in Artegna

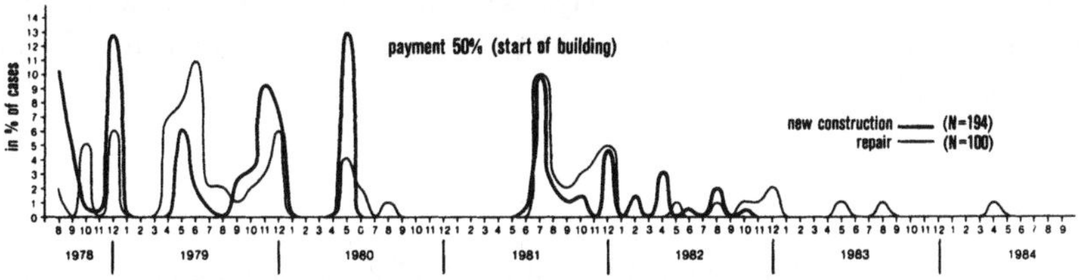

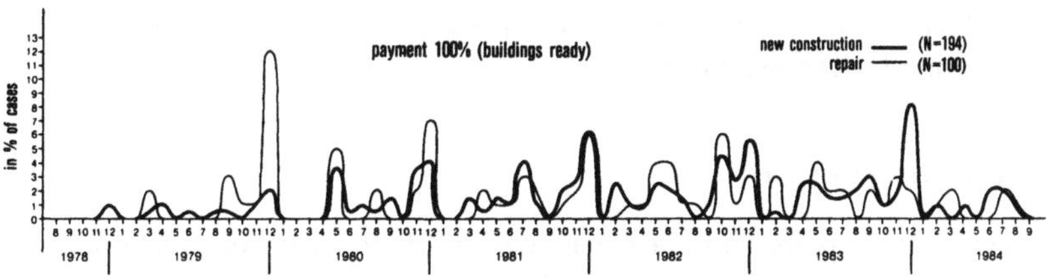

Source: Regione Autonoma Friuli - Venezia Giulia. Relazione sullo stato delle attività regionali per la ricostruzione delle zone colpite dai sismi del 1976. Trieste 1985. Allegato C

Fig. IV.10
Reconstruction cycle in Guatemala

LONG - TERM RECONSTRUCTION PATTERN

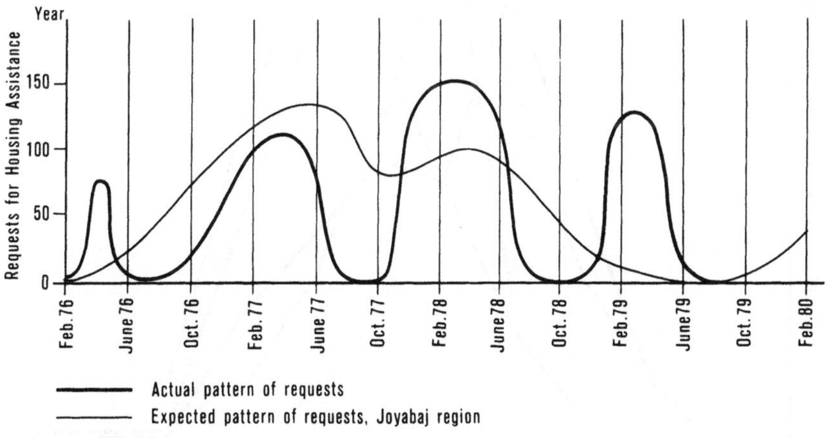

Courtesy: Cuny. 1983; p. 43

Degree of Destruction

The 45 communities of the category "destroyed" were subdivided into three groups according to degree of damage (see Fig. V.1) and private activities in repairing (law 30/77) and new building (law 63/77), expressed in 1977 lire, were assessed.

Repairs

Figure IV.12 shows that in the 18 most badly affected communities, compensation for repairs was drawn on much earlier than in the remaining 27 communities. Repairs were started and finished more quickly. Through 1979/1980, 20% more money for repairs was withdrawn than in the other categories. Altogether the curves are similar, with a steep take-off, a clear peak in 1979, and a smooth decrease until 1968.

New construction

Although the repairs show a clear peak in 1979, new construction is more evenly distributed between 1979 and 1982 (Fig. IV.13).

Fig. IV.11
Start of construction in sample communes

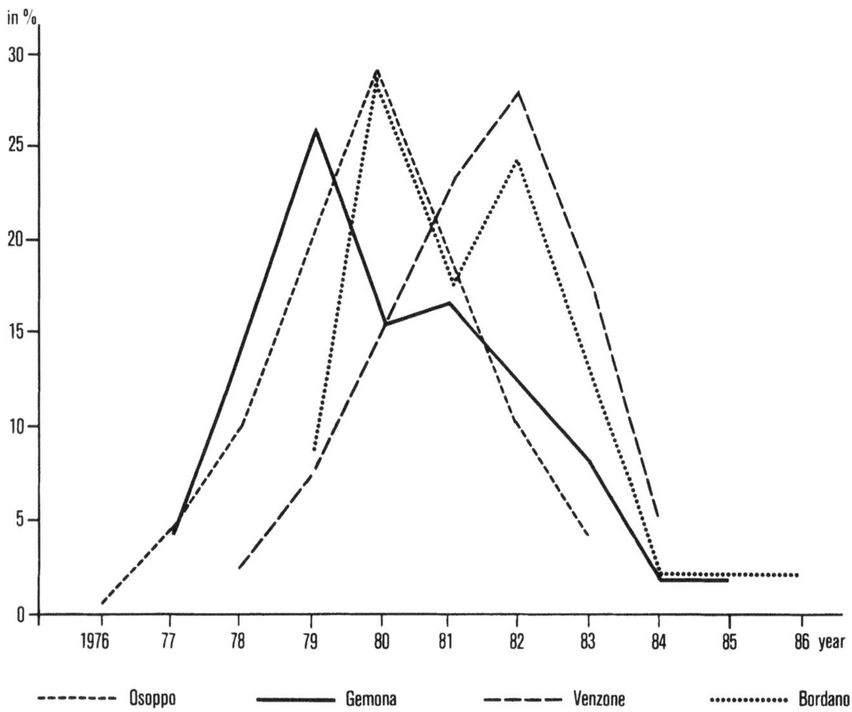

Source: own results

Fig. IV.12
Repairs: percentage of resources applied yearly, by extent of damage

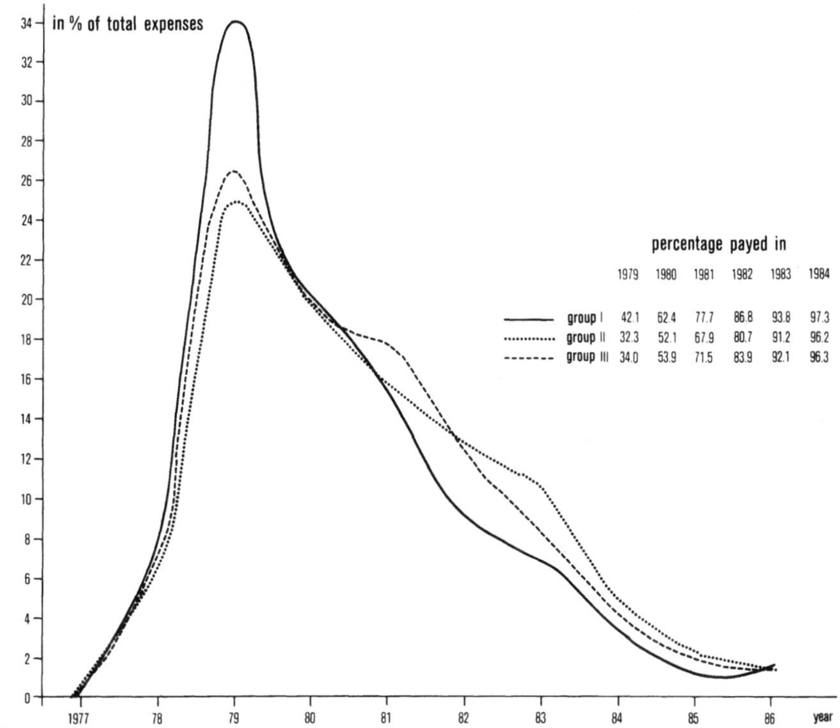

Courtesy: Segreteria Generale Straordinaria Udine 1987

As in the case of repairs, the most badly struck communities withdrew their compensation somewhat earlier than the rest of the communities. They have a distinct peak in subsidy payments in 1980 (one year later than the peak of the repairs). The total process therefore moved somewhat more quickly in them than in the communities of the second category, which are one year behind.

The last 14 comparatively less destroyed communities do not show the "normal distribution" with a distinctive peak, but stay on an even level between 1979 and 1981, even though they started more quickly because of less planning delay, e.g., when no redrawing of property lines was necessary.

In summary, despite bigger problems and/or greater destruction, building activities started earlier in the more destroyed communities, and rebuilding in them always reached its peak sooner than in the less affected ones. Evidently, private activities intensify according to the magnitude and gravity of problems.

4. Socioeconomic Development

This categorization of the 45 destroyed communities was derived from the cluster analysis described in Chapter VI.3. The 45 destroyed communities belong to clusters II–V, and we will look into their repair and building activities.

Repairs

Figure IV.14 with the four clusters II–V shows that three clusters (II–IV)—among them II with the best socioeconomic conditions—are rather similar, with a peak in repairs around 1979. Cluster III is fastest in calling off compensation money.

Cluster V, however, the communities with the worst socioeconomic conditions, differs visibly from the other clusters. Communities within cluster V start repairs on a lower level in 1979 and stay on this low level until 1983. Whereas clusters II–IV have called off three quarters of their compensation money by 1981, the amount in cluster V is only one half. Repairs are slower in the communities of economically retarded development.

Fig. IV.13
New construction: percentage of resources applied yearly

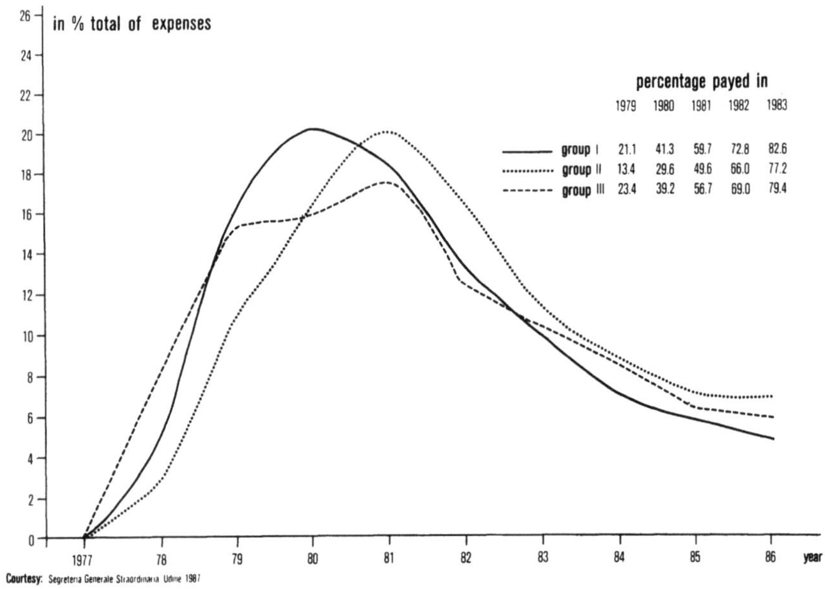

Courtesy: Segreteria Generale Straordinaria Udine 1987

New Construction

The difference just mentioned in repairs does not show up between cluster V and the other three (Fig. IV.15) when it comes to new construction.

Peaks differ between 1980 and 1981, and there is a surprising second peak for cluster V in 1983. But reconstruction in the form of new buildings proceeded rather evenly. There were no communities that suddenly expanded while others were given up for good. This was prevented by the framework of laws, although in view of regional planning such a result might not always have been desirable.

5. Differentiation of the Stages and Influential Factors

For repairs and new construction we can distinguish different stages in respect to time and intensity both for the whole disaster area and within this area.

Fig. IV.14

Repairs: percentage of resources applied yearly, by socio-economic type of commune

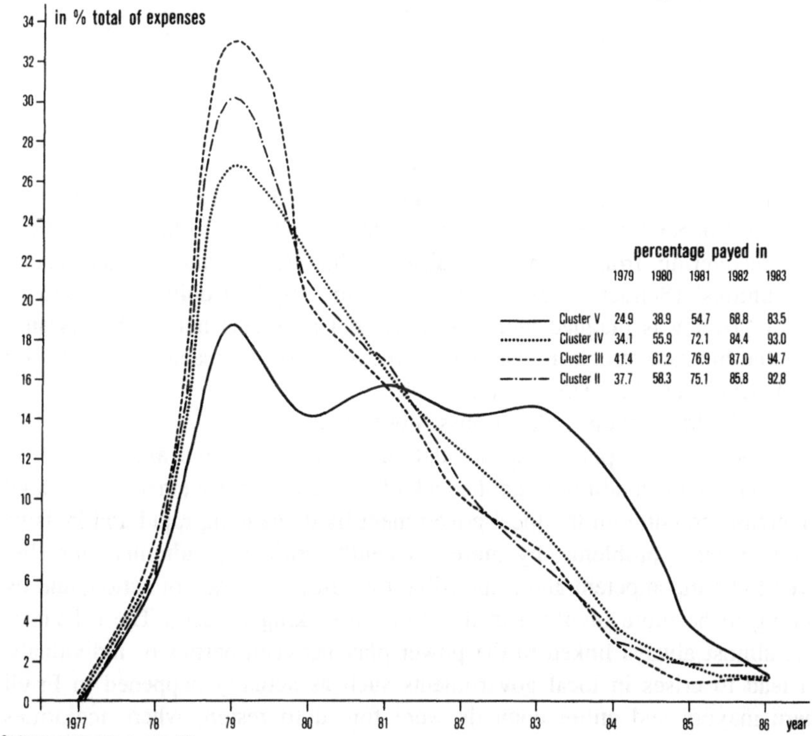

Courtesy: Segreteria Generale Straordinaria Udine 1987

Within the area both cycles are influenced by:

- the degree of destruction
- size or degree of socioeconomic development of a community, regenerative capacity, or marginal conditions revealed by cluster analysis (cf. p. 144ff.).

Bigger communities, when experiencing a similar degree of destruction, have to grapple with extra administrative and bureaucratic procedures, and these technical problems hamper a prompt payment of contributions more than in smaller communities because it takes longer to work through the many applications. In smaller communities, on the other hand, we do not find problems with the administrative procedures but with the personal qualifications of personnel and the social composition of the population that may be unable, for example, to cope with red tape. Socioeconomic background and size of community being the same, the degree of destruction has a uniform influence on the stages of reconstruction.

Thus, Rubin et al.'s influential factors "extent of damage" and "availability of resources" (our cluster groups) can be distinguished both in the repair and in the new construction cycles. The study of cycles and of their influential factors is important because they allow differentiation among communities within a vast disaster area and reveal how hazard management can take proper account of time and space variation. We wish to contribute, through such observations concerning the construction cycle, to an individualization of aid measures, verifying statements such as this one of Rubin et al.'s: "Small towns with few financial resources and limited staffs are likely to have more difficulty recovering from a massive disaster than larger, more sophisticated communities" (Scituate, Massachusetts, Blizzard 1978) (Rubin et al., 1985, p. 92), or "Hull was plagued with organizational problems before the disaster, when the blizzard struck, it exacerbated them" (Hull, Massachusetts, Blizzard 1978) (Rubin et al., 1985, p. 102).

As to building cycles, Cattarinussi observed:

"In the recovery phase, the population often exasperated and pressed by precarious living conditions and the lack of shortterm improvement, could tend to increase pressure on the local government by demanding rapid and incisive action to solve problems, by more insistently criticizing administrators believed to be incompetent and responsible for delays and lack of action, and by wanting to be more involved in the decision-making process. These behaviours, almost always linked to the power play between parties or individuals, can lead to crises in local governments such as actually happened in Friuli where mayors and entire councils were forced to resign, where majorities shifted, where commissioners from the Prefecture took over" (Cattarinussi and Tellia, 1978, p. 250, quoted after Drabek, 1986, p. 293).

6. Costs of Reconstruction

In the following table, Dobler, a member of our research team, refers to Di Sopra's estimate of damage. Although subsequently alleged to have been overestimated, it formed the starting point for all compensations.

Fabbro, too, claims 4000 to 4400 billion lire as the damage of the two earthquakes, 2000 billion alone in housing and infrastructure and 1000 billion for industry. Since within 10 years the lira because of inflation lost half of its value, exact data are difficult to obtain if they are not given with a fixed date. These global statements forming the initial estimate of damage influenced all policies since 1977. They had a political function because the compensation provided by the Italian government was based on these figures. Table IV.1 brings them together with the later expenditures.

For the new construction of an apartment or a house, a family of five persons in 1977 got an allowance of ca. 30 million lire (then approximately $50,000). This contribution grew by 1985 to more than 100 million lire (ca. $90,000). Such an allowance, on the average, covered half of the building costs. According to Fabbro and to data from our four test communities, this amounted to an average allowance of 49.2% to 49.5%.

For repairs the allowance was somewhat lower (42.3%). Relations between repair costs and costs of new construction therefore were between 1:1.5 and 1:2. Hazard research postulates that compensations must be adequate: "Assistance is often provided according to the principle that its beneficial effects are proportionate to the size of the budget. What is frequently overlooked is that this is true only to a point, and that assistance beyond this point tends to be counterproductive" (Klintenberg, 1979, p. 64).

Too small allowances do not mobilize private demand and private funds. When too high, allowances find their way into unnecessary, costly and unprofitable facilities (town halls, etc.) We tested the official data against the results from our survey in four selected communities, which will be presented later. Here we want to discuss them in a more general framework. There is noearthquake insurance in Italy (cf. the case of Newcastle).

Figure IV.16 shows the distribution of the allowances to private households. On the average, people in Gemona got 47%, in Osoppo 46.4%, in Venzone 58.4%, and in Bordano 60.3%. But if we look into details, the picture is much more different (Fig. IV.17). Size and equipment of residences, house types, townhouses versus detached houses, possibility of participating in the work on one's own house, starting time of construction (before or after public intervention to bring down building costs), or family size may explain some of the

Table IV.1 Earthquake damage according to Di Sopra

Damage for May earthquake		With 20–30% additional for September			Expenditures of the Regional Government, incl. 1984
($1 1976 = 842 Lire[a])	In bill. lire	In mil. US $	20%	30%	In bill. lire
Damage to productive facilities	1256.2	2111.18	1507.4	1633.1	ca. 750
Direct damage to	335.4	563.53			
Agriculture	202.6	340.59			
Industry	95.7	160.59			
Handicraft	17.8	30.00			
Trade / services	19.3	32.35			
Secondary damage by	920.8	1547.65			
Curtailment of production	608.9	1023.53			
Effects on income	112.0	188.24			
Interregional implications	199.9	335.88			
Damage of housing system	2018.9	3392.94	2422.7	2624.6	894
Damage to housing	1314.8	2209.41			
Destruction of houses	397.3	667.65			
Damage to dwellings	689.3	1158.24			
Clean-up costs	171.1	287.65			
Damage to outbuildings	41.0	68.82			
Damage to furnishings	16.1	27.06			
Damage to public facilities	704.2	1183.53			
Communal technical infrastructure	48.1	80.59			
Communal social infrastructure	279.2	469.41			
Regional infrastructure	66.9	112.35			
Cultural properties	310.0	521.18			
Hydrologic damage	145.0	243.53	174	188.5	ca. 87
Total	3420.1	5,748.24	4104.1	4446.2	1984: ca. 1731

Source: Di Sopra, 1977; Dobler, 1980, p. 61.
[a]$1 1976 = 842 Lire; $1 1987 = 1297 lire.

differences between the communities. Also important is the availability of private savings. Eighty percent of all families in Osoppo could draw on their own savings against 62% in Gemona, 53% in Bordano, and only 43% in Venzone. Families in Osoppo could obtain loans from relatives (17.1%) or from banks (44.3% of households questioned) more easily than families in other communities (Venzone: 4.5% and 13.6%). Evidently, the economic wealth of Osoppo, an industrial village with a steel plant, booming because of the high demand for engineering steel for construction, is manifested in the savings activities and credit rating of its citizens.

7. Regional Differentiation of Reconstruction Costs

In principle, we have to assume that every repair or new construction applied for led to an apartment or a house, so that the number of registered public interventions should equal the number of reconstructed flats or buildings. Let us compare the figures for two communities, Osoppo and Trasaghis.

	Osoppo		Trasaghis	
Situation in May 1986 (reported by community)				
Repairs	334		335	
		= 961		= 1271
New construction	627		936	
Situation in Fall 1984 (reported by region)				
Applications for repairs	304		282	
		= 899		= 1164
Applications for new construction	595		882	

Concerning these two communities, the figures are comparable, although derived from various sources. But within Friuli the expenses of the region for private reconstruction differed widely per community and per capita in relation to degree of destruction, size of family/household, and number of interventions by the State.

The following paragraph deals with the expenditures of the region in constant lire (as of 1977). To assess the value in the late 1980s, the figures mentioned must be multiplied at least by four because of inflation. Per resident citizen, between 1.1 million lire (in Tolmezzo and Spilimbergo) and 13.1

Fig. IV.15
New construction: percentage of resources applied yearly, by socio-economic type of commune

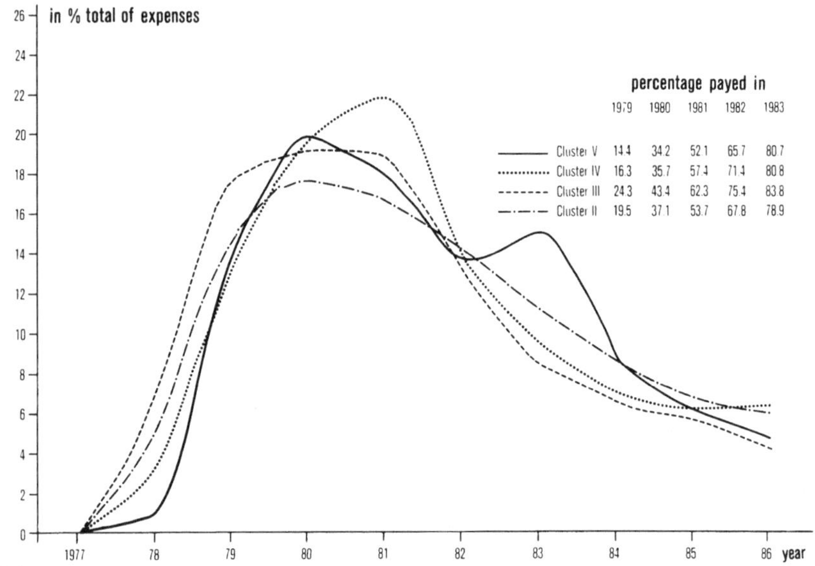

Courtesy: Segreteria Generale Straordinaria Udine 1987

Fig. IV.16
Percentage of total costs covered by subsidies

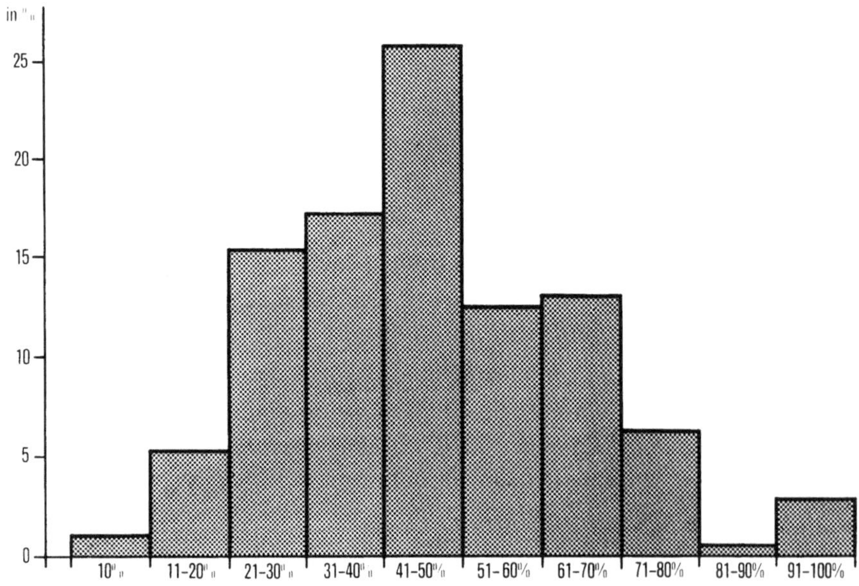

Source: own results

million lire (in Bordano) were transferred for private reconstruction to the concerned communities.

1. Degree of Destruction and Reconstruction Costs

The higher the proportion of newbuilt residences, the higher the expenditures via subsidies. If we divide the 45 destroyed communities following the degree of damage into six groups, the following amount of money per capita was transferred on average to the communities:

Table IV.2 Subsidies for private reconstruction per capita, according to extent of damage

Group I	More than 80 points heavily destroyed (10 communities)	6.2 million lire
Group II	70–79 points (8 communities)	6.2 million lire
Group III	60–69 points (6 communities)	6.2 million lire
Group IV	50–59 points (7 communities)	4.3 million lire
Group V	40–49 points (7 communities)	3.6 million lire
Group VI	Less than 40 points (7 communities)	1.5 million lire

It is true, of course, that with greater destruction the number of newly built (not repaired) houses and of subsidies increases. But there are large differences despite an equal degree of destruction. Local circumstances and decisions evidently must have influenced the type of reconstruction. In Montenars, for instance, three quarters of activities went into new construction; in Amaro, on the other hand, three quarters went into repairs, although both communities belong to group I with more than 80 points on the destruction index. Evidently, the fact that Amaro belongs to the seismic risk zone of Tolmezzo, where some seismic safety building regulations already were in effect before 1976, made it possible simply to repair more houses than elsewhere.

2. *Family Size and Reconstruction Costs*

The smaller a family or household was, the higher the subsidies per single intervention. One-person households received money for 67.5m^2 of living space, two-person households 82.5m^2, three-person households 95.5m^2 , and so forth. In the small, remote mountain communities with over-aged population and many one- or two-person households, much more intervention per capita had to be organized than elsewhere.

3. *Number of Interventions and Reconstruction Costs*

On average, for every family concerned one building intervention was made by the authorities. There was a possibility to get subsidies also for two or three houses, but they were reduced for the second and third house, and therefore many people entitled to such a subsidy did not make use of it. On the other hand, many older persons forfeited their subsidies because they did not claim them and decided to stay (with title deed) in their prefabs. This might level out some of the discrepancies.

But in many mountain communities many more interventions took place than families/households were registered present. Registered inhabitants are always more in number than those really living in these communities with high out-migration. In Bordano, 392 repairs and 311 new constructions, altogether 703 public interventions, took place, where the current population amounted to 827 persons in 1984. On average, one building intervention was made for 1.2 persons, while the average family size was 2.5 persons (census 1981). Clauzetto announced 396 repairs and 120 new constructions (516 interventions) for a population of 624 current persons, bringing about one activity for 1.2 persons, while on average the family size was 2.3 persons.

If we compare the figures of average family size (1981) with the number of persons per activity, we can clearly perceive that in the periphery of Friuli many more interventions took place than families and/or households really existed (Fig. IV.18). This is true even if we go back to the census figures of 1975, because mountain communities in the meantime lost some of their inhabitants. In the category of the communities with the highest degree of destruction (Venzone, Osoppo, Gemona, and Resiutta) one public intervention was made for 2.0 to 2.7 persons (average family size: 2.3 to 2.7 persons). In the mountain communities, however, the number of public interventions equaled nearly double the figure of families and/or households present (see the darkest sectors in Fig. IV.18)

In many mountain areas not all public interventions may have brought about a "visible" house, even if residential space had grown disproportionally. Before the earthquake Bordano had 413 houses, 129 of which were empty, 284 occupied. Today we find here 317 new and 50 repaired houses, in total 367, which would house 918 persons. But only 827 live here. Thirty seven homes (10%) must be empty. Here as in many other communities, residential space has been created far beyond the state before the disaster. The predicted decrease of population between 10% and 30% for the years to come will keep many of the new houses empty, if migration from other parts of Italy, especially the Mezzogiorno, does not fill them. This would lead to political unrest because of the ethnic homogeneity of the Friulians.

The regional differentiation of reconstruction costs may be resumed as follows:

- Expenditures per capita for private reconstruction as well as for public facilities in the mountain areas were much higher than in the more centrally located communities.

- In the mountain communities many more building interventions took place and subsidies were spent than houses built. Evidently through subsidies the previous building stock was enlarged, improved, and embellished. Two flats were turned into three. This can be explained by the high dissipation of titles and ownerships in an area with high out-migration.

- Despite a high population decrease, residential space was created, which is the more unnecessary, the more peripheral the locations chosen for rebuilding.

- Despite enormous subsidies that were given to these communities, out-migration was not prevented, although perhaps retarded for a short period.

One possiblity that might have reduced these undesired consequences of a law that earmarked money to be spent in the place of loss only could have been the transfer of subsidies mentioned above. But this came too late and was granted only reluctantly because the mayors wanted to "keep their flock together."

The last figure of this chapter (Fig. IV.19) indicates communities that received especially high or low subsidies.

8. *Change Following Disaster: A Review*

Hazard research has heretofore tried to discover change following disaster through measurement applied by outsiders. Before reporting our findings based on the questioning of insiders, we should present a short review of the state of the art (Buttimer, 1979). Drabek (1986, p. 183) postulates that "centralized responses tended to occur in smaller communities. The largest cities tended to cooperate badly, and centralized responses were nonexistent." Three models of recovery are hypothesized: "1) autonomous, 2) kinship and 3) institutional reconstruction" (Drabek, 1986, p. 280).

The models in Friuli would be (1) private (autonomous plus kinship), (2) cooperative, and (3) public with different degrees of participation of the victims and corresponding acceptance of these models. A typical result is also that all energy to rebuild is consumed by the construction of one's own house, and that engagement for the community as such, for the environment and its quality, is low.

This new home becomes the Moloch that consumes all energy, savings, thoughts, and aspirations. There are many findings to show that after a disaster the newly built residences show a much higher standard of size and equipment than before. "Many of the homes that were rebuilt are larger and have a higher assessed value than the previous structures . . . the owners were able to get low interest loans that allowed them to build previously unaffordable luxury features. . ."(Rubin et al., 1985, p. 91).

Most households therefore were highly indebted. While all energy concentrates on the home and the new house, the former settlement structure dissolves. Built-up areas expand, because everyone wishes a freestanding house to walk around. ". . . the town is not compact but sprawls in clusters on separate hills" (Germen, 1978, p. 71 on New Gediz). Lamping (1986, p. 5) gives another example from Turkey: ". . . The monotony of the new settlements is in sharp contrast to the compact form of the old traditional villages."

Fabbro (1985, p. 31–32, Fig. IV.20) postulates three patterns in overcoming a disaster. He thinks that Friuli fits into the third type, where a dualistic spatial structure is forming. To some extent two villages are lying side by side, representing past and future, traditional heritage, and uniform adaption to the automotive era, inhabited by residents different in age and in economic condition, and respectively village-centered or outside-oriented. Because they received their subsidies, hired building gangs, and found architects at different times they followed dissimilar prevailing taste, encouraged by architectural journals and imposed on them by still faster neighbors.

Citizens' participation may be a troublesome, time–consuming, and expensive game, and one not always successful. But if the authorities plan and build regardless of the ideas of the victims, as Lamping finds in Turkey and Guatemala, the results are nearly always the same: aversion to the new houses and out-migration from them. Without consulting the victims, project ruins will be produced with which the population cannot build up identification. As long as the costs and the bills for maintenance are paid by the sponsors, the population grudgingly dwells in these buildings. When further payments are cancelled, their houses built by "someone at the top" will soon be abandoned.

Public participation is vitally important for the implementation of policies and strategies" (Quingkang, 1982, p. 205). These results of hazard research will now be compared with the findings of one of our team members, A. Bardola, in her study of four communities.

Fig. IV.17
Apportionment of subsidy ratio, by commune

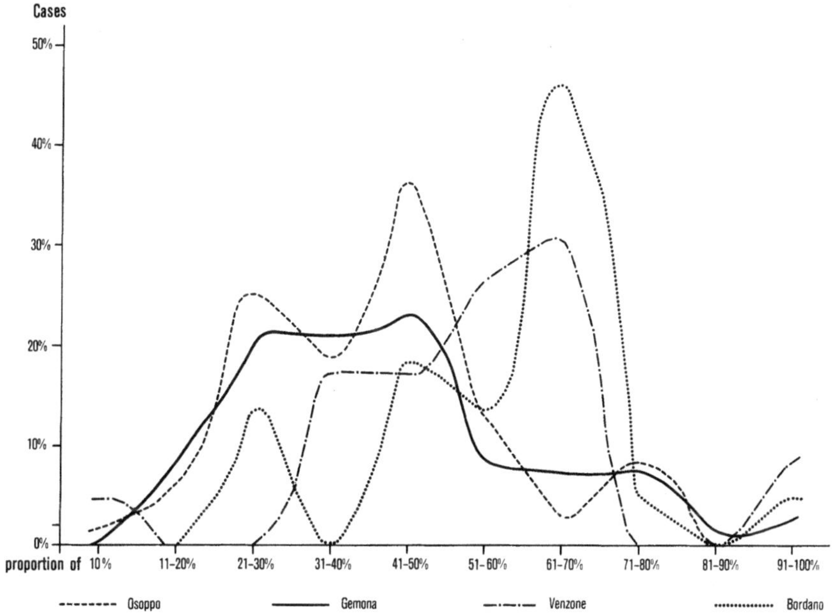

Source: own results

Fig. IV.18

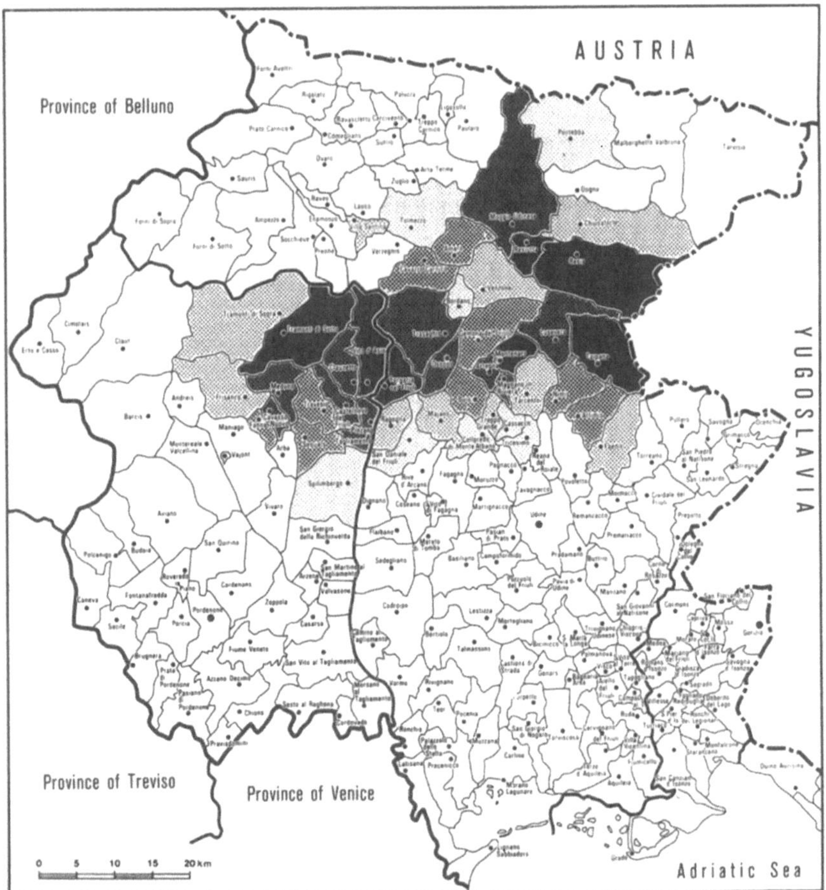

Ratio of number of families to number of applications for assistance, individual communes, 1981

- \> 0 (7 communities)
- 0.1 - 0.5 (11 communities)
- 0.6 - 0.9 (9 communities)
- 1.0 - 2.5 (10 communities)
- \> 2.6 (8 communities)

Courtesy: Regione Autonoma Friuli - Venezia Giulia. Relazione sullo stato delle attività regionali per la ricostruzione delle zone colpite dai sismi del 1976. Trieste 1985. p. 26. Allegato B

Fig. IV.19

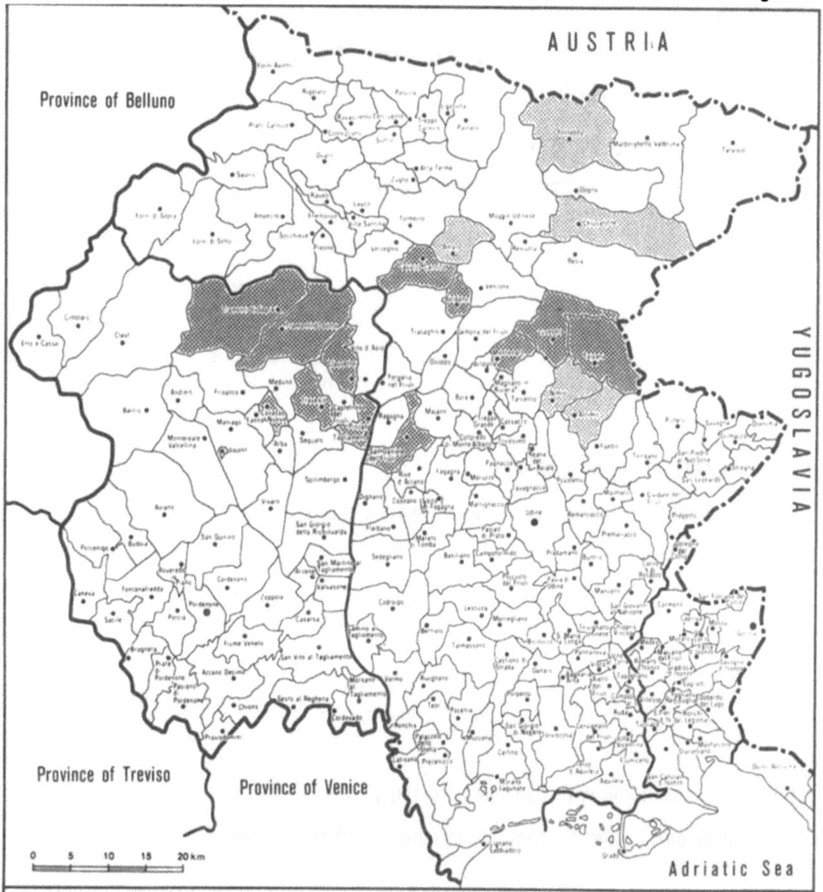

Communes receiving resources above and below the mean

- at least 25% more than means of group, according to extent of damage
- at least 25% less than means of group, according to extent of damage

Courtesy: Regione Autonoma Friuli - Venezia Giulia. Relazione sullo stato delle attività regionali per la ricostruzione delle zone colpite dai sismi del 1976. Trieste 1985. p. 26. Allegato B

Fig. IV.20

Scheme showing to extent of damage the three different ways communes coped with catastrophe

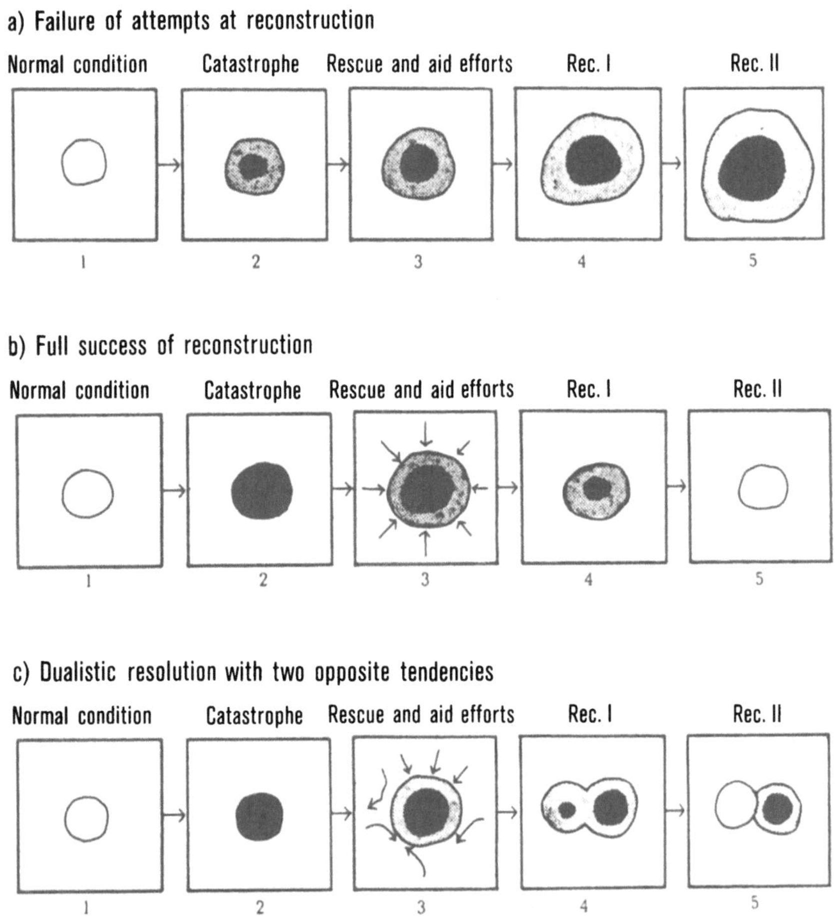

○ success ● failure

Courtesy:
Fabbro, S., 1985, Fig. 2–4, p. 31/32

V
Reconstruction in Four Communities: A Case Study

Chapter IV aggregated the cycles of the reconstruction story within an area of 4800 km^2 with 150,000 victims and losses amounting to \$6.3 billion in the perspective of the financial planning of the two provinces Udine and Pordenone. Of course these stages are not reflected evenly in the different communities, and even their curves mirror only insufficiently the picture of hopes and disappointments, hardships, and stress of the individual families. Therefore, Anne Bardola, a member of our research team, studied more detailed reconstruction in four selected communities (Osoppo, Gemona, Venzone, and Bordano).

This case study should contribute to answering the following questions:

1. How did reconstruction change the residential and settlement pattern?
2. What was the assessment of the victims in regard to the strategies of reconstruction?
3. How did the victims evaluate their residential situation in comparison to their situation before the earthquake?

The case study started out with unstructured interviews of experts from which a questionnaire was developed. The investigation took place in October 1986 in the four communities mentioned. The rate of return amounted to 72% (n = 284). The questionnaire gave information about 883 persons belonging to the 284 households.

The four test communities were selected with the intention of giving examples for different strategies and results of reconstruction, and comparing them. Regarding the framework of the comparison, the communities complied with the following requirements:

- they were all located in the center of the disaster area
- they were all within the category heavily destroyed (more than 85 points) (Fig. V.1)
- they were located along the line of development from Tolmezzo to Udine.

As to their socioeconomic development, however (Fig. V.2) the four selected communities represented four different types that, according to Fabbro (1985, p. 156ff.) can be characterized as follows:

Type A (= Osoppo). Economically well developed community, striving upward with high attraction for its hinterland; progressive industrialization process with initial development of public and private tertiary sector. These communities expanded and diversified their local labor market during the last 10 years. They have a stable demographic state and a growing population. They are attractive for in-migrants or emigrants returning from abroad.

Type B (= Gemona). Communities participating in the economic development, but without originating any impulses on their own. Communities of a certain size, which in the past showed relatively good economic growth but stagnate now.

Type C (= Venzone). Communities at the periphery of the developing areas. Unstable in respect to socioeconomic development, with negative symptoms or tendencies toward deterioration. Weak signs of improvement through tourism. Structurally backward with few work places and strong dependency from active communities.

Type D (= Bordano). Communities in peripheries with negative economic and social trends and in the vicinity of active neighbor-communities, which benefited from widespread investment in rebuilding after the disaster but could not create jobs. This development is connected to a loss of structure and a change of functions within the settlement pattern.

1. The Situation Before the Disaster

a) Osoppo (1985: 2639 Inhabitants)

Osoppo is a middle-sized community in the hill area of Friuli on the left bank of the Tagliamento river. It is divided into the main village of Osoppo (with the industrial zone of Rivoli) and the small fraction of Pineta. The immediate proximity of a freeway exit and the favorable location with respect to Public Highway 463 provides Osoppo with excellent access to all important traffic arteries

Until 1970 Osoppo was a semirural center with about 190 farms, most of them part-time. By the middle of the 1970s the number of farmholds had diminished rapidly. This decline was caused mainly by the rapid industrialization of the Industrial Zone Rivoli Osoppo, 2 km south of the main village, founded in 1962 and originating with the initiative of a group of private entrepreneurs (Pittini steelworks). Situated between the market centers of Gemona and Buia, Osoppo would have remained a place of out-migration without this industrial zone. But this stopped the further decrease in population, and by creating jobs even led to an increase in population.

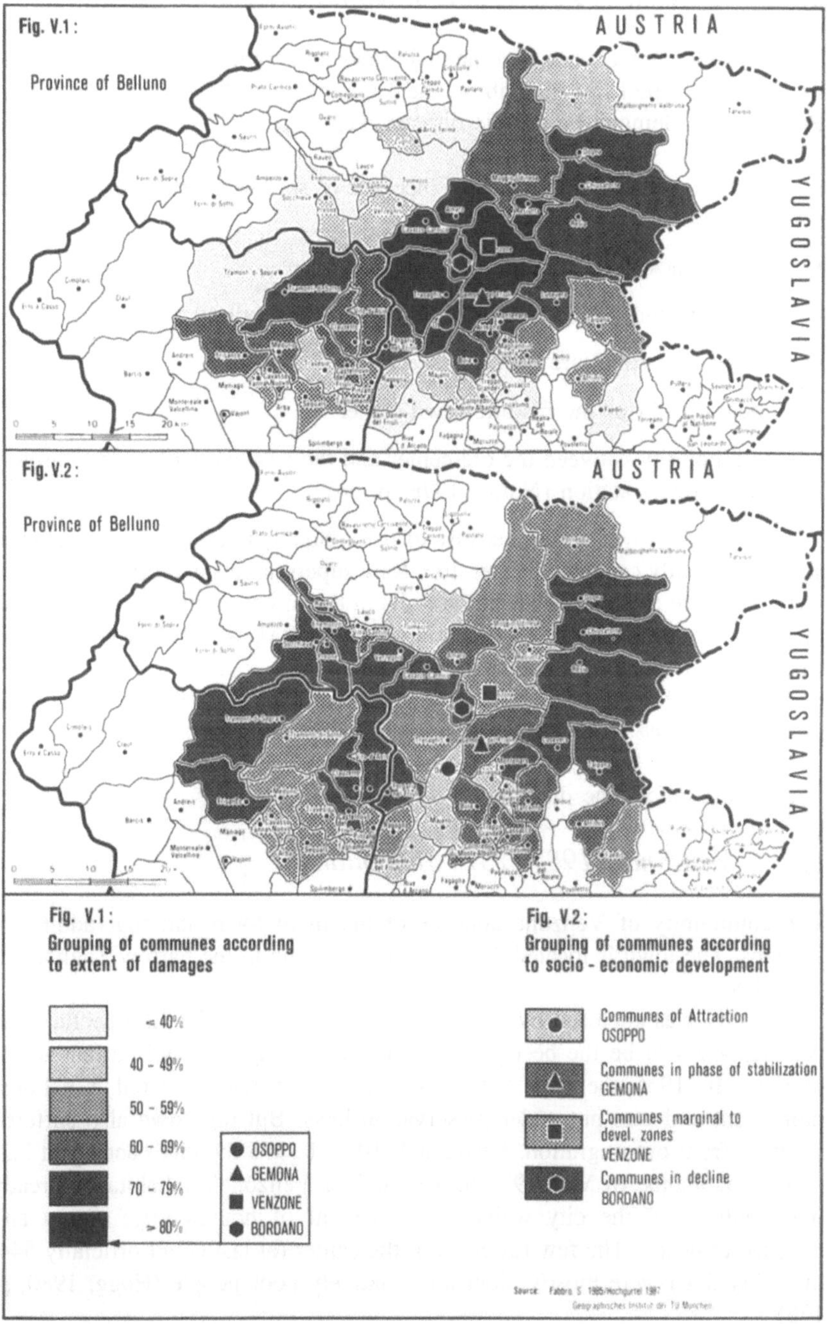
Fig. V.1:
Province of Belluno
AUSTRIA
YUGOSLAVIA
Fig. V.2:
Province of Belluno
AUSTRIA
YUGOSLAVIA
Fig. V.1:
Grouping of communes according to extent of damages
< 40%
40 - 49%
50 - 59%
60 - 69%
70 - 79%
> 80%
OSOPPO
GEMONA
VENZONE
BORDANO
Fig. V.2:
Grouping of communes according to socio - economic development
Communes of Attraction OSOPPO
Communes in phase of stabilization GEMONA
Communes marginal to devel. zones VENZONE
Communes in decline BORDANO

b) Gemona (1985: 11,065 Inhabitants)

The city of Gemona consists of the main town and eight sections. The city with its historical center lies high above the Tagliamento valley on an alluvial fan. Problems of Gemona before the disaster were:

- a dispersed settlement pattern (because of serious shortcomings in urban policies)
- partly lacking public utilities (sewage, etc.)
- crisis in the tertiary sector (commerce and services)
- crisis of agriculture
- degradation of the historic town center
- a bipolarity between the old center and the new parts of the city around the railway station (Nimis, 1978, p. 34).

Shops in the town center, because of the antiquated building structure, were small and hardly capable of extension. New supermarkets and other businesses were therefore established at the foothills in the direction of the station and at the crossings of Highway 13 (Austria–Adriatic Sea), where they could profit from the through traffic and tourism (Nimis, 1978, p. 50).

Already before the earthquake the historical center of Gemona was occupied by only 50% of its former population, as shown by a cartographic survey by the architect Nimis, who had been commissioned with an urban renewal project just before the disaster.

c) Venzone: (1985: 2374 Inhabitants)

The community of Venzone consists of the main town and four additional sections. The town is located 30 km north of Udine in the narrow Tagliamento valley.

The city was enclosed by a town wall and a moat in the 13th century and was supposed to be the best preserved medieval city of Friuli, attractive for tourism. In 1965 the area within the city walls was declared a national monument and was put under preservation laws. But this town also suffered strongly from out-migration. Between 1910 and 1975 the city center lost half of its inhabitants. In May 1976 nearly 80% of Venzone's inhabitants already lived outside of the city walls. Forty percent of houses were empty and exposed to decay. The few residents of the center (at last count officially 544, in reality 370) were mostly older and relatively poor people (Hogg, 1980, p. 175).

d) Bordano: (1985: 842 inhabitants)

The community consists of the main village of Bordano and a smaller section (Interneppo). It was characteristic for this community that:

- a relatively high proportion of inhabitants declared "residents" had their permanent residence elsewhere
- there was a high proportion of old people (1981: 26% 60 years and older); in 1986, 75 of the 104 singles were older than 60 years
- There was a high proportion of empty houses
- the parceling out of property titles was extremely high
- there were no jobs besides part-time agriculture on very small farm-holds
- it occupied a remote location on the right bank of the Tagliamento, squeezed between the flood area of the river and the rockfall threat from the steep mountains.

Photo 3: Main Street in Osoppo with new arcades.

2. Reconstruction Policies

a) *Osoppo*

Osoppo, ruled by a coalition of socialists (PSI) and communists (PCI), had its population concentrated in a huge baraccopolis outside of the village. This facilitated the clearing of debris, which was so thorough that one could speak of a "tabula rasa." Most owners could not even find the remnants of their house walls to fix the former boundary lines. The preexisting network of roads, however, although smoothed out and enlarged, was supposed to be retained if possible, in order to provide a possibility of preserving the historical layout of neighborhoods and hence family links and social connections. The new Osoppo was not to be a slavish copy of the past but a concept considering present requisites without losing the peculiarities of the village and its links with tradition (cf. *Osoppo*, 1986, p. 15). The greatest planning problem was to arrive at an agreement among the property owners over the redrawing of land boundaries. This consumed a great deal of time and caused many difficulties. The historical center was, in the long run, successfully rebuilt privately, although following a master plan with exact specifications as to size and height of buildings. Altogether, planning in Osoppo was rather prosaic and technocratic, without any special design virtues. The main street is seamed by arcades, which look rather traditional but never before existed (Photo 3).

b) *Gemona*

The suggestion not to rebuild the old city center of Gemona would have raised the question of what to do with an area of more than 60 ha. This would also have resulted in huge claims for compensation. Furthermore, seismic risk estimates pointed to the middle of the alluvial fan, the location of the old center, as one of the least dangerous places. And of course the old town also meant the essence of the identity not only of the Gemonese people but also of the whole surrounding population. By reconstruction of the cathedral, city hall (Photo 4), and the old city, the planners hoped to prevent the further "sliding downhill" of Gemona to the plains.

On the other hand, the planners recognized the difficulties of the center's revitalization. Nimis (1976, p. 160) pointed to the fact that aspiration and admiration of the people were attracted by the new houses down in the plains: modern, clean, and comfortable. Therefore he pleaded for a strong initiative to be focused on the historic center, carrying the expense of building permit limitations in the outskirts. The ruling local Christian Democrats (DC) agreed with these recommendations, but then fell back into a laissez-faire style, thus causing serious problems for the center's reanimation. The historical center was reconstructed with considerable fidelity to its traditional structure and partly as a copy of its original appearance (Via Bini, Palazzo Communale). The city core near the city hall was rebuilt through public intervention, the restprivately and by cooperatives (Photos 4 and 5).

Photo 4: City hall of Gemona, built ca. 1500, destroyed 1976 and reconstructed in historical style.

Photo 5: Bank in Gemona, in close neighborhood to city hall.

c) Venzone

The victims in Venzone opted at first for a "free reconstruction" through private initiative, which would have been possible because of the poorly defined legal status of a national monument. In the face of great difficulties (high dispersion of land ownership, owners who could not be located) on one hand, and the possibility of additional financing on the other, a great majority of owners voted for a historical reconstruction, but they were hardly aware of the duration and costs of such a type of rebuilding. The community administration too (a coalition of socialists, communists, and social democrats [Psdi]) favored this concept. In December 1977 a decree was issued by the Italian state, which

Photo 6: Front gate of a palazzo in Venzone, restored stone by stone.

Photo 7: The reconstructed city wall of Venzone divides the community into an inner part of museum character and an individualistic outer part.

Photo 8: City hall of Bordano, a community with strong out migration and 824 inhabitants left. Background: new terraced houses.

put Venzone under full preservation as an ancient monument and ordered its historical reconstruction. The main reason for this kind of rebuilding (accord ing to the architect in charge, a member of the Ministry for the Preservation of Ancient Monuments) was the declared intention of the citizens of Venzone to do so (Messaggero Veneto, May 6, 1986, p. 4). The reconstruction of the old town of Venzone was carried through as a perfect historical rebuilding "stone by stone" including the medieval wall and moat (Photo 6). This was possible because plans existed predating the disaster. The whole reconstruction was enacted by the Ministry for the Preservation of Ancient Monuments and the community of Venzone in the form of public intervention. The area was divided into 18 building blocks. Participation of the owners was restricted to the free choice of an architect for their block. In the beginning it was difficult to obtain a building license for construction in the area outside of the walls (Photo 7). But here also private initiative triumphed to a notable extent.

d) Bordano

For some time people considered that Bordano should not be rebuilt in the same place because under the shock of the second earthquake many victims expressed their wish to move elsewhere. Bordano should cross the Tagliamento and become a new section of Osoppo, an expanding town with good prospects for the future. But only one third of the citizens of Bordano agreed to this plan. The community area of Bordano would have offered the possiblity to extend the settlement into the alluvial land of the Tagliamento and comply with the desire of many inhabitants for a free-standing house. But the reflec tions of the mayor in charge during the planning period pointed toward the preservation of agriculturally usable land. Part–time farming would provide an additional income for a village without any workplaces. A more extended built-up area would also have dictated more costly utilities. And only part of the population might have been able to profit from such an individualistic type of reconstruction. Only one lot measured 600 m^2, two or three were bigger than 300 m^2; 90% of all land ownership amounted to building lots between 70 and 90 m^2. Because of co-ownership an individual might possess only a couple of square meters. Reconstruction planning therefore would compromise between the predisaster situation and free–standing houses. To preserve the former dense spacing and to save space as well as provide everyone with an individual property and their own piece of garden, row houses seemed most appropriate. In a series of meetings, models of reconstruction were discussed (Photo 8).

Finally row houses were agreed to because no realistic counterproposals could be made. According to the mayor they were most appropriate in respect to boundaries, sizes of lots, and so forth. Opposition centered on objections to the "model of expropriation and row houses" and disappointment was great that not everybody was provided with a free–standing house. Because of these reservations many citizens lacked the willingness and determination to deal with problems (and opportunities) of this model of rebuilding.

In total, two thirds of the village was built by public intervention in the form of modern row houses. Some people ironically called them "i treni" (the railway trains).

The administration was blamed for their "bossy" methods (Tentori, 1986, p. 109). Indeed, of the four communities under comparison Bordano suffered the most radical change. While in the three other communities political majorities remained stable, Bordano's reconstruction took place under three different local governments. The elections of 1980, before rebuilding even had begun, saw a coalition of christian democrats (DC) and social democrats (Psdi) win against the communists (PCI), who had held office during the planning period. In 1985 a "free movement" with members from all parties won over the DC-Psdi coalition.

These events correspond with the results of our inquiry insofar that in Bordano alone the number of those who would not build the same way again as they did predominates slightly. In Osoppo, in contrast, three quarters of the interviewees agreed with their former decisions. The other two communities hold positions between these extremes. A chi-square test of significant differences in the evaluation of reconstruction, as among communities that fol-

Photo 9: "I treni"—the "railway trains" of Bordano (compare with Photo 14!).

lowed different policies, showed a contingency co-efficient of 0.22. Especially in Bordano the evaluation of the reconstruction reveals shortcomings.

This report on organizational types of reconstruction and the communities' policies allow us to hypothesize different outcomes. To document these in the view of the *insiders* (the victims) (Buttimer, 1979) was the issue of our survey, which asked the following questions:

- How was the victims' assessment of different policies during reconstruction?
- Are the people who were affected content with the possibilities of participation offered to them?
- Are there certain groups that are particularly content with reconstruction; how is the assessment of different age groups?
- How is the evaluation of the residential situation today in comparison with the predisaster situation?
- Did the social climate in the community change under the impact of reconstruction?

3. Changes in Residential Structures and Settlement Patterns

Our survey stated that the average residential space of the interviewees amounted to 120 m^2 per household after reconstruction. Only 17% of units had fewer than 80 m^2.

Our survey supplements the two last census figures.

During reconstruction there was a visible shift toward more space per household. This increase is strongest in communities with positive trends (Fabbro, 1985, p. 86).

The whole sample had an average living space per capita of more than 35 m^2. The assumption that public building was particularly generous can be verified by our results: the space is almost as large as that of private reconstruction. Public reconstruction in Bordano was somewhat more modest, however, with 92 m^2 on average per home (Venzone 121, Gemona 109). But in consideration of the smaller population, Bordano makes up with an average living space of 35.4 m^2 in comparison to Venzone with 35.5 and Gemona with 41.8. Private reconstruction in Venzone and Bordano is comparatively modest, with 95 m^2 and 102 m^2 respectively (total private average 129.4 m^2).

Our interviews showed that private rebuilders in both communities had a higher influence on building than in Gemona or Osoppo. This, and a propor-

Table V.1 Average size of dwelling (by community)

Community	Census 1971	Census 1981	1986 interviews
Bordano	60.5 m^2	89.9 m^2	97.0 m^2
Venzone	78.4 m^2	99.4 m^2	112.4 m^2
Gemona	90.1 m^2	108.5 m^2	116.6 m^2
Osoppo	92.4 m^2	104.3 m^2	147.2 m^2

Table V.2 Average size of home by intervention

Intervention	Average size	Average number of inhabitants	Space/person
Cooperative	106.9 m^2	3.56	30.0 m^2
Public	107.4 m^2	2.85	37.7 m^2
Private	129.4 m^2	3.37	38.4 m^2

tion above average doing their own work within this group (Bordano 28%) could explain the high rate of compensations in these villages. In Bordano and Venzone private builders started later with reconstruction. They could go back to partly normalized building costs after 1982.

A comparison of living space between the historical centers and the outskirts shows that the average share of middle–sized residences (81–140 m^2.) is always bigger in the outskirts-group. The share of big residences (141–200 m^2) is highest in the centers, in contrast to the fact that bigger and still growing households predominantly lived outside of the historical centers. Family size of four or more persons prevailed, with 56% in the outskirts in contrast to 40% in the center.

Evidently two principles of rebuilding were in conflict. On one hand, compensations were fitted mainly to needs (family size). On the other side it seemed important to build "as it had looked before." This meant for the historical centers that residential space, already half empty before the disaster, for restoration's sake was rebuilt, and many private persons were forced by the zoning plan to build wastefully large. One extreme case amounted to 600 m^2, 400 of them to remain empty. At the outskirts of towns and villages private builders obtained adequate space whereas in the centers too many and too big residences were built for old couples and singles. Because of excessive costs,

Fig. V.3
Distribution of dwelling size in the four sample communes

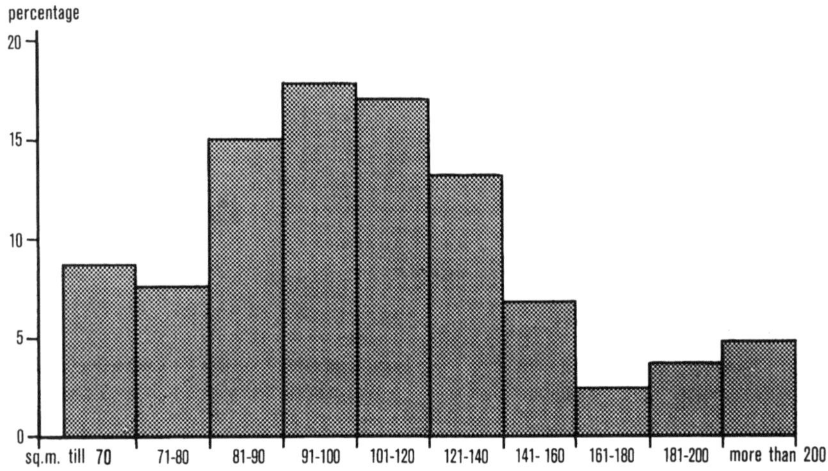

the owners or former owners were not capable of paying for "their" homes, which were forced on them by the administration in search of a museum character for the city core.

Fabbro (1985, p. 87) states, in respect to the development of residential space, that despite a decrease of population of 2.6% in the last decade, reconstruction brought an increase far beyond the real needs of the population. Reconstruction nearly doubled residential space: instead of one room per capita there are now two of them, one at public cost and one at private (i.e., out of savings or loans). Fabbro (1985, p. 21) therefore speaks of a "double reconstruction."

We found a retired emigrant couple who had come back after many years in South Africa. Because they were entitled to compensation, the community asked them to join a group of seven families in the same situation. They were offered a lot in a group of eight units of terraced houses. In 1988 only two were occupied. The rest of the terraced houses had been finished only in the rough, because the other owners built only as long as the compensation lasted. They did not need these houses because their main residence was somewhere else. They participated in this program only in order not to forfeit their claims. Now they speculate that after the five years of a freeze on sales, someone will buy their semifinished buildings so that they can change a compensation title into cash.

Fig. V.4
Distribution of dwelling area by commune

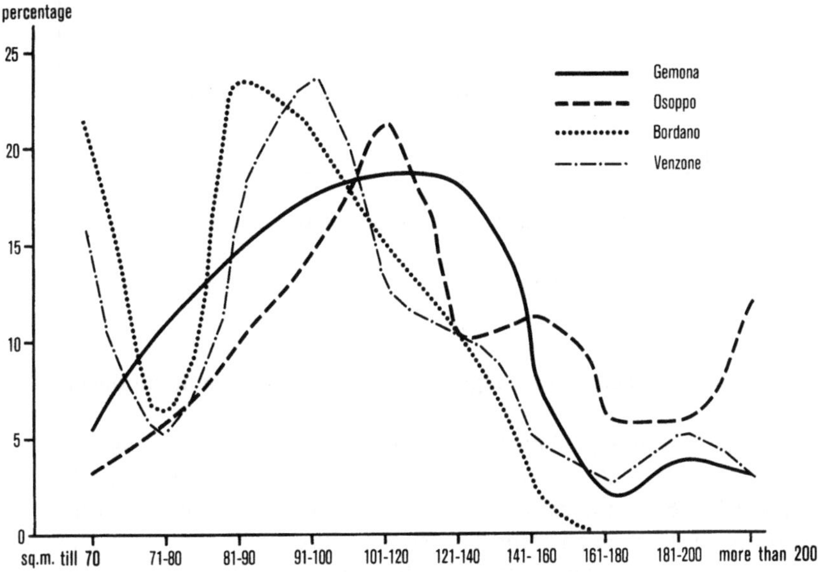

Fig. V.5
Distribution of dwelling area by "intervento" type of builder

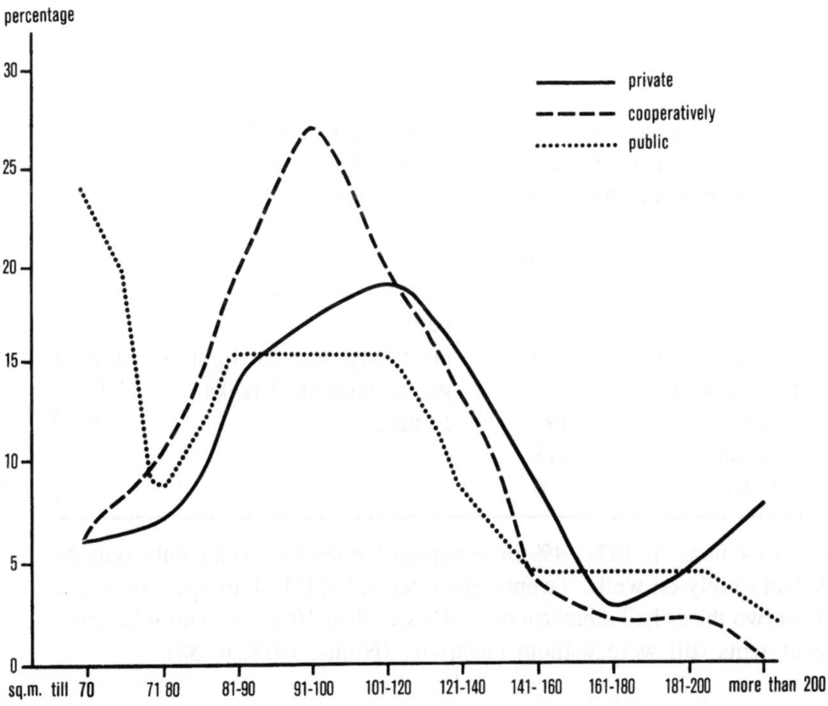

Table V.3 Development of housing stock in Bordano

Year	Stock	Thereof empty	Occupied houses
1961	451	83 (18.4%)	368
1971	424	107 (25.2%)	317 (= 326 dwellings/ISTAT 1971)
1976	415	132 (31.8%)	283

The small village of Bordano is another example of the building orgy:

buildings in 1976: 415	void 132	occupied 283
buildings in 1986: 367	newly built 317	repaired 50

Today 830 persons live here, on average 2.26 persons for one building operation.

As early as 1978, it was predicted that reconstruction would be excessively overdone: on May 31, 9666 families consisting of 48,446 persons lived in prefabs, of which 2469 "families" consisted of one person only. But 18,000 applications for reconstruction had already been submitted. Even if we take into account that some of the people concerned had found another place to live, among these 18,000 applications must have been many aiming at the reconstruction of empty houses or second homes.

4. The Improvement of Housing Stock

Reconstruction led to an enormous growth in housing quality, which without the earthquake would not have been materialized in decades. The census of 1971 described Bordano with the following figures.

The 326 occupied apartments were equipped with:

water	269	bathroom	79
water outside the flat	49	electricity only for lighting	178
water from a well	8	gas in individual tanks	326
toilets inside	192	furnace	6
toilets outside	218		
without toilet	6		

Even in Gemona in 1971, 4% of occupied homes had water only outside and 5% had to rely on wells. Twenty eight percent still had an open air toilet, fewer than two thirds had a bathroom, and fewer than 10% had central heating. Five apartments still were without electricity (Nimis, 1978, p. 58).

Photo 10: New buildings in Bordano

Today all apartments are equipped in conformity to modern standards, from tiled bathrooms to central heating. People agree that with regard to standards of equipment, reconstruction meant a leap from medieval to modern times (Photo 10).

The growth in quality is acknowledged by the residents. Our interviewees agreed that in a comparison between past and present the new houses were brighter, airier, and more practical and convenient (Fig. V.15).

5. Changes in the Structure of Accommodations

The houses in Friuli were mostly products of local culture. They were a mirror of the Friulian society, which in an optimal way had regulated its environmental relations. But the influx of financial means from other regions to such an enormous extent led to the import of many alien influences on social and cultural traits. They made Friuli "less Friulian" (Barbina, 1977, p. 24). The predisaster building stock in urban centers was of mainly medieval origin and up to 70% consisted of houses built in traditional styles. Out of these 70%

Fig. V.6

Mean dwelling area per private unit in 1971 und 1981, according to extent of damage within the 45 destroyed communities

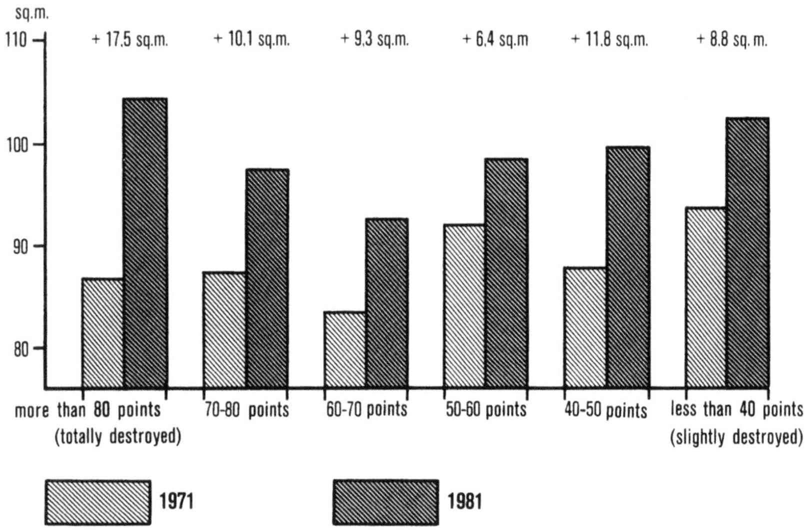

Source: 11 Censimento generale della popolazione. ISTAT. Roma 1973
12 Censimento generale della popolazione. ISTAT. Roma 1984

Fig. V.7

Traditional Settlement pattern

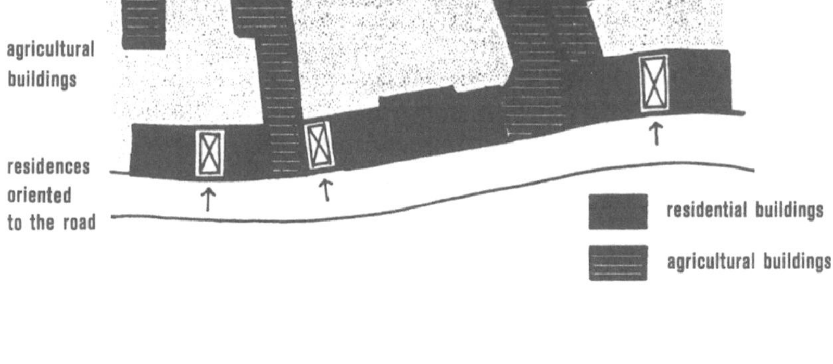

Photo 11: A cluster of farms built side by side illustrating traditional neighborhood-relations.

Photo 12: Individual new buildings in the outskirts of Venzone.

three quarters were combined with agricultural activities (Fabbro, 1985, p. 47).

The random sample showed that in our four communities under survey the proportion of single–standing houses before the earthquake averaged between 21% and 24%. Predominant in all villages were one– and two–family houses built wall to wall.

In the bigger communities of Gemona and Venzone, however, there was a relatively higher proportion, more than 20%, of multiple-family dwellings, indicating the old urban building stock with big multistory houses. This proportion of multiple-family dwellings (+4%) has diminished after the disaster in all cities except in Bordano. The free-standing one-family type house in our sample has increased by 46%. Fabbro (1985, p. 49) also claims that the change in accommodation stock went in the direction of free–standing one–family homes. This happened at the expense of one– and two–family houses built wall to wall along the roads and farmyards, which were reduced from family homes. This happened at the expense of one and two-family houses 56.5% to 35%. Their arrangement and type of construction were well fitted to the needs of an agrarian life style and consisted of house, stable, sheds, inner court, and garden, as shown in Figure V.7 and Photo 11.

The strongest decrease of this house type happened in Bordano and Osoppo (here especially in respect to the former common farmyards). It was replaced by row houses with individual entrance and small gardens, mostly separated from each other by concrete walls (Photo 12). Row houses increased in Osoppo from zero to 23%, in Bordano from zero to 37%. In all four sample communities modern row houses, practically nonexistent before the earthquake, increased to 16%.

Cooperative as well as public reconstruction laid stress on row houses and multiple–family dwellings, but private individual rebuilding preferred the free–standing one–family house (46%). In a social perspective, this development was traceable to the aspiring and dynamic groups, which manifested competition and ambition to demonstrate status, shown in the preference for modern architectural trends (Fabbro, 1985, p. 84). In our sample the highest percentage of single family houses (31.5%) belonged to the second age-group (35–45 years), the group for which such homes showed the greatest increase (+ 18%). To gain this end, these people had to raise the highest loans and renounce the solidarity principle of mutual aid (cooperatives); they now represent the very group that shows highest satisfaction with reconstruction (see Table V.8).

6. Changes in the Location of Residences

Of our interviewees, 48.4% lived in the same part of their village as before the disaster, 23.8% in its vicinity, and 26.4% somewhere else.

The highest "as before"-rate of 68.9% existed in Venzone followed by Gemona (54.3%). Taken together with the second category, "in the vicinity," the smallest change of sites took place in Venzone and Bordano (80%).

The proportion of those now living "somewhere else" is remarkably high in Osoppo, with 40% compared with Gemona (24%), Bordano (20%) and Venzone (18%). The explanation is the forced reallocation of all private parcels through the community administration that took place in Osoppo's center, so that 18% of these 40% of "somewhere else" reflect site shifts in the center.

Older people are more likely to live at the same site as before (Table V.4). The age group 36 to 45 years has the highest share of "elsewhere," with 50.7%.

Table V.4 Present compared with former site of dwelling, by age in percent

Site	Unchanged	Nearby	Elsewhere	Total
Age				
25-36 years	53.8%	26.9%	19.2% (07.0%)	26
36-45 years	25.4%	23.9%	50.7% (47.9%)	67
46-55 years	50.0%	30.6%	19.4% (16.9%)	62
56-65 years	61.8%	23.5%	14.7% (14.1%)	68
older than 65 years	65.2%	13.0%	21.7% (14.1%)	46

Of our interviewees 47.2% live in Gemona's, Osoppo's or Venzone's historical center. Half of those (49.5%) who now live in the outskirts did not live either there or nearby before. This means that more of of them came from the historic center. It was mostly young people who moved out (Table V.5).

Table V.5 Dwelling site in town center and outside, in percent

Age	Now outside (93)	Both at present and previously elsewhere (46)
25-35 years	47.8%	8.7%
36-45 years	61.4%	5.0%
46-55 years	42.9%	21.7%
56-65 years	38.5%	13.0%
older than 65 years	15.4%	6.5%

In the second age group (36 - 45 years) 61.4% live outside of the historic center today. In the other age groups the percentage of those living in the center is considerably higher. With the exception of the first age group, it is the rule that the older they are the more people tend to live in the historic center. According to these figures and in view of the many empty apartments in the centers, the predisaster trend of superannuation may well continue. Reconstruction has intensified the segregation of age groups.

If we correlate the present site with the general satisfaction with living conditions the following relation is the result.

Table V.6 Dwelling site and satisfaction with conditions, in percent

	Less content	Rather content	Very content	n
Center	19.8%	35.7%	44.4%	(126)
Outside	59.0%	22.6%	65.6%	(93)

Significance: 0.0081; 2 degrees of freedom.

Therefore, contrary to the widely propagated slogan "dov'era e com'era" (where it was and how it was before) (Fabbro, 1985, p. 81), the earthquake was for nearly every second person a welcome opportunity to realize their dream of a one–family house outside of the historical center. Reasons for thei emigration are manifold. On one hand it is the continuation and intensification of previously existing trends, and the earthquake was the welcome occasion

Photo 13: The ideal of the Friulians, too, is a house with garden, freestanding bungalow type. Reconstruction made it possible in the outskirts of Venzone.

especially for those who sooner or later would have moved anyway. On the other hand the earthquake was the immediate cause for internal and intercommunal mobility, because many of those impatient to rebuild did not like to endure long periods of uncertainty, waiting, inactivity, and the small range of participation permitted owners within the area of preservation of ancient monuments by the administration. Whoever was able to went out and built by himself (Photo 13).

Even the laws supported this trend, because anyone who had already agreed to public intervention and for whom the community already vicariously had built in the center could get his compensation money back to invest it elsewhere. Of course he thereby forfeited his claim on his former property. But now the community had to provide a new owner or renter for this abandoned property in the center. The communities therefore had to find renters or buyers for residential space without corresponding demand.

Where single houses stood merely scattered before, the space in between has been filled up in the meantime. In Gemona a stream of buildings has grown downhill toward the railway and beyond. Fabbro (1983, p. 28) in 1982 found that 30% of all building permits issued related to sites that had not been rebuilt before the disaster. Partly those were sites that had been explicitly exempted from use for building before, in order to save valuable agricultural soils. This "sliding down to the plains" was supposed to have been prevented through the ambitious reconstruction of the old town. But this policy failed as it did in Venzone, where only physical conditions restricted a vaster expansion. Today it is difficult to find citizens to fill up Gemona's and Venzone's historic centers.

Following the computations of Nimis (1978, p. 227), Gemona's historic center was planned for 3400 persons, double the figure of inhabitants before the disaster (in 1976, 1532). The higher standard of the new and bigger apartments and the good supply of services should be sufficient to stop a future dispersion of population and to realize a maximum of urban concentration. But this idea included a strict denial of building licenses outside of the center. A weak administration could not insist on these restrictions. When we asked our interviewees in 1986, fewer than 1000 were already living in the historic center, and 2000 were still missing.

The dissipation of the settlement, although it did not conform with the planners' purposes, reflects the requirements of the population. It wants its share of the amenities propagated by architectural journals and media all over the Western countries. The highest status is derived from a bungalow with garden, terrace, balcony, and garages. According to these aspirations our interviewees criticized the lack of green space and playgrounds, balconies, and gardens that a historic reconstruction had denied the center. A traditional museum-like layout is not adequate for modern needs, they stated. To quantify such statements we asked people who had rebuilt in and outside of the center to what degree they would do so again.

Osoppo clearly differs from the two towns that were rebuilt in a historical style.

Also, the assessment of the new house and its environment on a semantic differential test reveals that the new house is valued more highly by those who rebuilt in the outskirts (Fig. V.11).

The quality of the present environment is valued more negatively than be-

Fig. V.8
New Settlement pattern

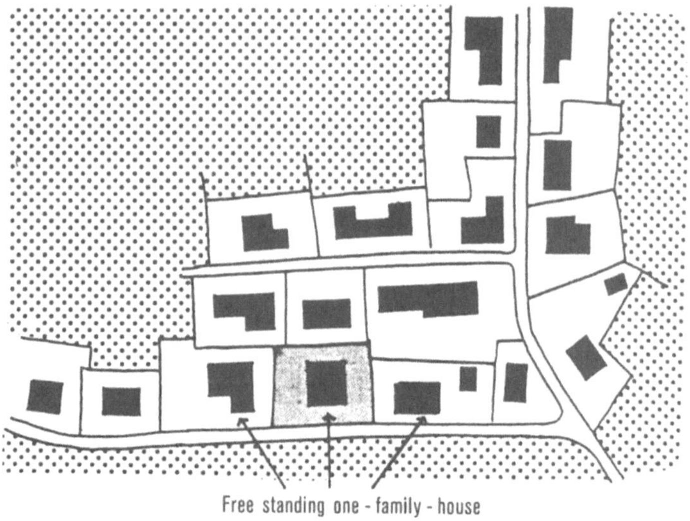

Courtesy: Fabbro, S., 1985, Fig. 5, p. 35

Table V.7 Percentage of positive answers to the question "Would you rebuild again as you did?"

	Gemona	Osoppo	Venzone
Center	48.5%	75.0%	47.6%
Outside	59.0%	73.3%	87.0%

Fig. V.9
Assessment of present environment by people living in the town center resp. those outside

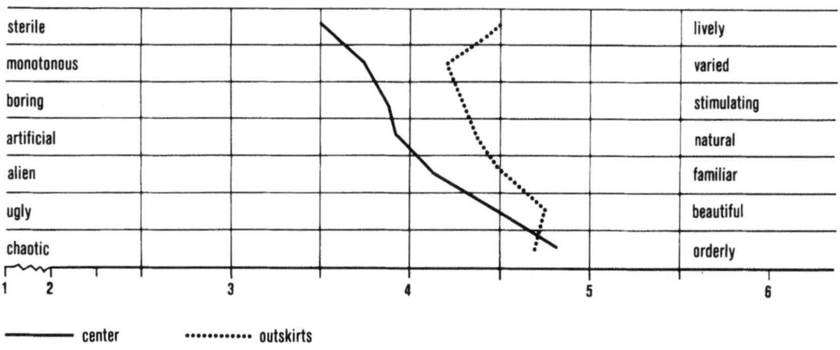

Fig. V.10
Assessement of former environment by people living in the town center resp. those outside

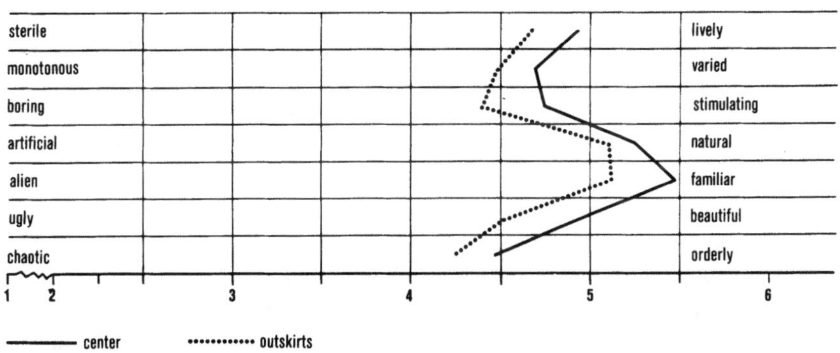

fore the earthquake. A main result of our inquiry is that during reconstruction the group that represents the most active part of the population used rebuilding to improve its situation. It left the former environment to create a new home for the family corresponding to its needs. The result, however, is ambiguous. This age group is more content with reconstruction and the present situation than others. An important aim of rebuilding seems attained. But this group also most strongly opposed the plans proposed, and thereby thwarted them. The vacuum in the old centers and the dispersion in the outskirts is mainly attributable to them.

Fig. V.11
Assessment of living space before and today

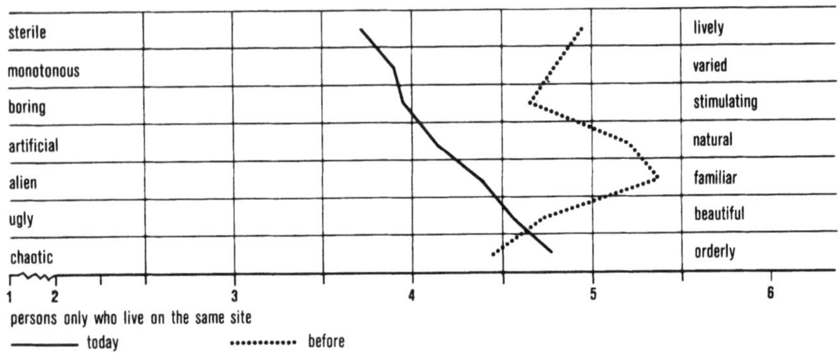

7. Changes in Property Titles

Friulians attribute a high value to real estate. "To own property has a primarily symbolic value for Friulians, not a capitalistic one. It is evidence, for liberation from the dependency in which the Friulian farmer was kept for centuries" (Nimis, 1978, p. 122). "The house is symbolic for everything that is close to our hearts: full independence, the liberty to speak, to vote, to live, to act, and to move freely: all this is represented in 'the house'," (Ronza, 1976, p. 67). Friulians agree that they are suffering from "the stone sickness." This fact in many respects has impeded decision making during reconstruction. "If reconstruction is not to be treated in a totally individualistic manner, we have to overcome this individualism in order to facilitate reconstruction against a collective social background. We will have to compromise between these traditionally antithetical ideologies" (Nimis, 1978, p. 122). Our interviews aimed at uncovering how far such collective control was accepted.

Our sample before the disaster showed the following types of control: own property or joint property, 81% , renters 15.6% , other 3.3%.

The high property rate was mainly caused by the agrarian structure of the region. Only Gemona, because of its size and urban architecture, showed a quota worth mentioning, of 25.7% renters. Rented apartments played no role in the other villages. By far the highest rate of former renters is to be found among those who rebuilt cooperatively (38.2%). Public intervention with almost 80%, has the highest proportion of former owners, and another 13% of

Fig. V.12

Plan of Bordano before the earthquake

Source: ricostruire 1980, 10/11, p.19

Fig. V.13

Plan of Bordano for reconstruction

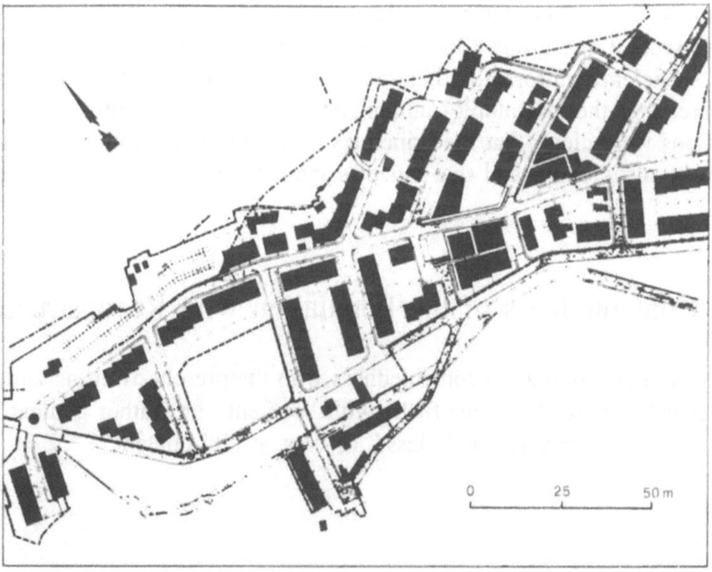

Source: ricostruire 1980, 10/11, p. 20

joint proprietors.

If we compare communities, Bordano has the most joint property, with 20% reflecting the high fragmentation of property before the earthquake. For reconstruction in Bordano the land had to be subdivided into 200 m^2 lots for public intervention (see Fig. V.13). This lead to a huge field reallotment during the process of public reconstruction, according to one planner.

In the meantime, control over land has shifted toward individual ownership with an increase of 31%. Comparing age groups, property titles shifted in favor of the second age group (36 to 45 years). In this group the propotion owning property was relatively low before (52.2%) and the rate of renters high (32.8%). The growth in property titles in this age group amounted to 40.3%, the highest rate of all groups.

This shift demonstrates thar reconstruction led not only to more and better dwelling space, but also restructured property rights. The generation of the young adults used reconstruction as a chance to profit from redistribution. Reconstruction with its complicated planning, requirement for initiative, and the reallotment of parcels, helped young families to ownership that normally would have been aquired by them much later, through saving or inheritance.

No wonder that the age group now 36 to 45 years and who were 24 to 33 years at the time of the disaster would to a high degree (two thirds), reconstruct today in the same way as they actually did. And in their general assessment of reconstruction also, this group expresses the highest positive evaluation, even if the differences between the age groups are not statistically significant. On the other hand, the very young ones are less enthusisatic in their statements, because they were too young and excluded from decision making in the period of the "big haul."

The second age group (36–45 years) more than others believes that they were sufficiently informed, that there were ample possibilities for private intiative, that the community administration did a good job, that laws and directions were clear, that the principles of "where and how" to rebuild were observed sufficiently, and that citizen's interests were respected in the planning.

8. Contentedness of the Population with Reconstruction

The overall question as to contentedness with the present dwelling situation is answered by more than four fifths with "content" or "rather content." Only some 16% are "only partly," "less," or "not at all" content.

Table V.8 Average assessment of reconstruction by age groups

1.	25-35 years	1.92
2.	36-45 years	2.05
3.	46-55 years	1.97
4.	56-65 years	1.98
5.	older than 65 years	1.87

Table V.9 Answers to the question of contentedness with present dwelling situation (house, neighbors, environment).

Content	54.0%
Rather content	30.6%
Partly content	7.5%
Less content	8.2%
Not at all content	0.7%

Therefore reconstruction in Friuli seems to be successful (Fig. V.14).

Fig. V.14

Opinions concerning reconstruction

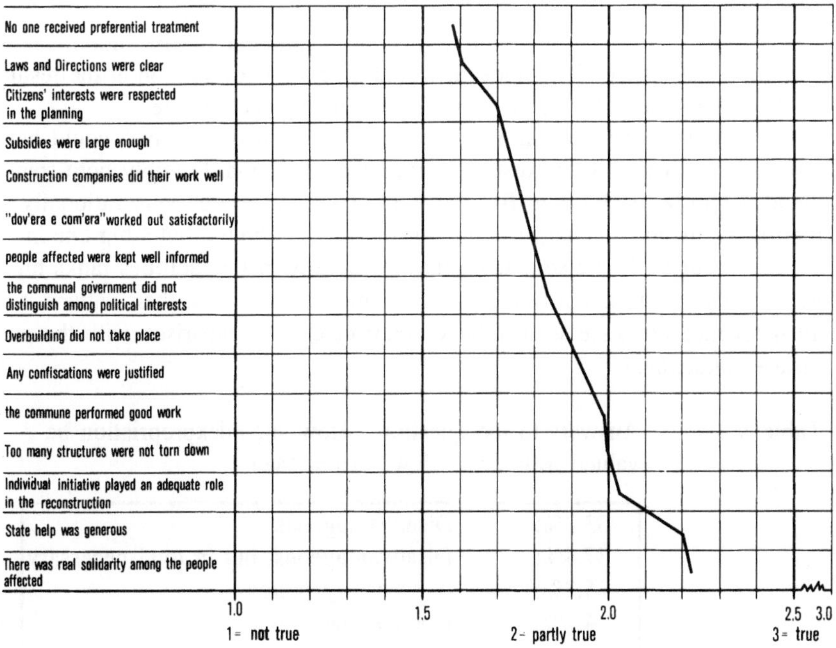

Of course, the use of the construct of "contentedness" in empirical social research is more dubious than many others. At any rate it is necessary to test this "contentedness" with some other indicators. Through including questions as to the attitude toward confiscation or the wish to alter plans we can at least control this doubtful construct of contentedness.

It has already been mentioned that house and real estate are important to Friulians, as we can presume from facts such as the high volume of investment in land and property ownership as well as from subjective evaluations of these attributes. Therefore they should be antagonistic toward all invasions into this domain. During the phase of reconstruction, Cattarinussi et al. (1981, p. 64) claimed that more than half of their interviewees were negative toward expropriation. The latter authors even believed that this was an astonishingly low rate.

In our investigation we found that more than half of our interviewees had been affected by expropriation, those closer to the centers more than others. This means that nearly two thirds, on second thought, believed that expropriation was necessary or even useful (Table V.10).

Inquiry in four communities, not altogether representative, therefore show in present perspective that expropriation—seen from the final positive outcome of the whole reconstruction—was also seen as positive. If 60% would reconstruct in the same way as they did, against 40% who would not, the dubious construct of "contentedness" is positively tested another time.

Another method was comparison of house and environment before and after the disaster. The semantic differential shows (Fig. V.15) that on average the present situation is evaluated as more positive than the previous one.

Of course, this could be interpreted as adaptation of the individual to cognitive dissonance. Who, indeed, would give a low evaluation of the results of the many years of hard work while rebuilding one's home?

On the other hand also a glorification of "the good old days" could have taken place, distorting the results in the other direction. Possibly both trends could neutralize each other. But our interviewees answered very rationally in their assessments: no change in the noisiness of the house, but a big one in its seismic security. It is also evident that the quality of the facilities ranks high, but that the environment (the sum of all individual houses, their pattern, and all other measures) are seen to have deteriorated in comparison with the pre-disaster situation.

Table V.10 Answers to the question, "How shouldexpropriation be valued in todays' view?" (n = 148)

33.3%:	Disadvantageous
37.4%:	Disadvantageous, but in total necessary
25.2%:	A necessary policy
4.1%:	Don't know

Fig. V.15

Comparison "then and now"

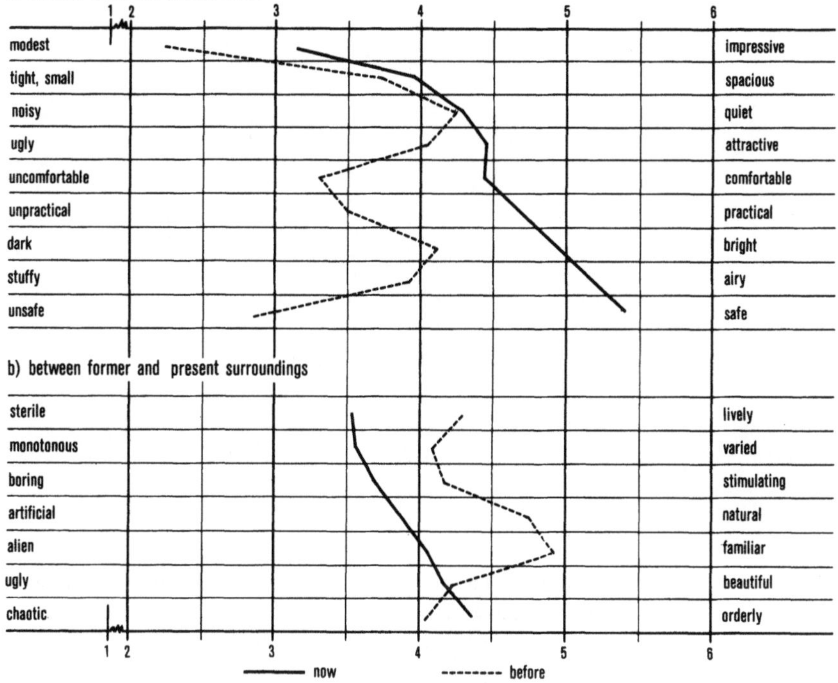

Fig. V.16

Assessment of Reconstruction, by commune

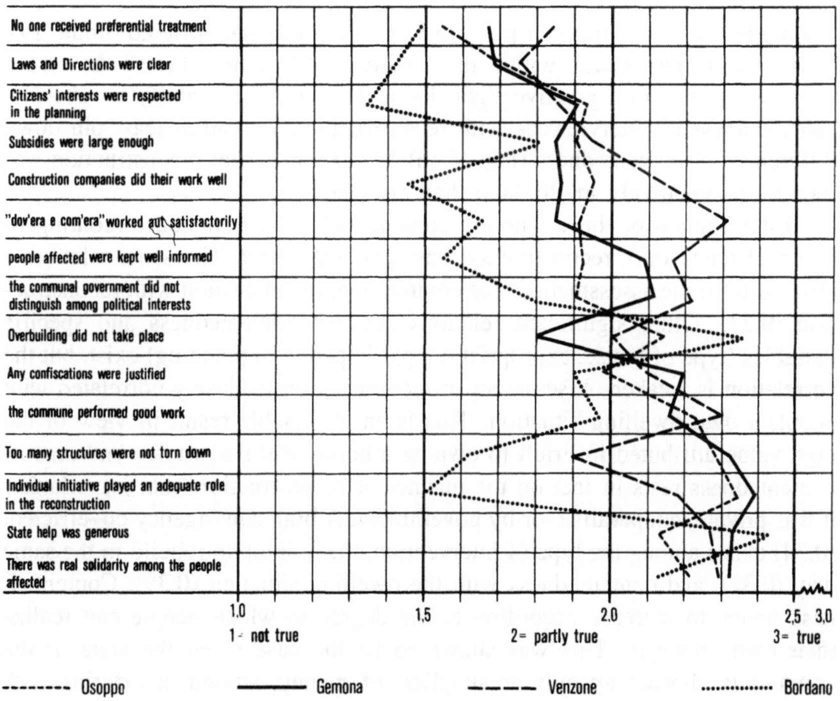

Photo 14: Row houses in the new village of Portis avoid the monotony of Bordano's "railway trains" and go back to traditional elements of local architecture (Architect: Roberto Pirzio Biroli).

This very difference between the assessment of one's new home and its embedding into the environment indicates that reconstruction planning might have been oriented too much at functional aims like clear property conditions, seismic security, sanitary equipment and a prevalent urban esthetic. The identity of the Friulians with their place of home evidently has suffered (photo 14).

Reviewing our results within the construct of "contentedness" the reconstruction is seen in a positive light. Even if we cannot quantify an absolute contentedness of interviewees with reconstruction, we can at least pin down differences of contentedness (Fig. V.16), for example, that reconstruction was seen more negatively in Bordano than elsewhere.

In the opinion of those who are content with their new house (contingency co–efficient 0.3), reconstruction has been successful. The same relation is also valid for the assessment of the environment in connection with reconstruction (0.27). Other significant relations between contentedness and specific variables (type of house, expropriation, participation in planning) exist, but the correlation is weaker. Assessment of reconstruction is closely correlated with people's own dwelling situation. This is an expectable result in view of the high value attributed in Friuli to owning a home. But the whole syndrome of contentedness rests in fact on the manner of reconstruction, that is, whether it was private, cooperative, or by government. A high contingency co–efficient (0.34) exists among the type of intervention, the opinion to rebuild in the same way (0.31), and contentedness with the dwelling situation (0.39). Contentedness seems to increase according to the degree to which people can realize their own concepts. This was shown to be the case when the state or the community showed up only as suppliers of money, without interfering with individual plans. Especially positive was the assessment by those who could realize their plans outside of the historical centers.

9. Change in Social Life

There is much debate in the literature over whether a disaster initiates social change. Some claim that this is the case (Bates et al., 1982), whereas others hypothesize that predisaster trends are speeded up in social behavior (Friesema et al., 1979, p. 87). In the emergency stage solidarity and harmony prevail, but later on old conflicts show up again and new ones are added (Quarantelli, 1977, p. 100). In Cattarinussi's and Strassoldo's survey in 1977, 81% of interviewees in Venzone claimed, that the social climate was ruled by patronage, fraud, meannesses of all kind, and envy. Partly there was an unscrupulous fight for better subsidies, and for a better starting position in the competition for funds. Such conflicts in the reconstruction period create a "monetarization of hitherto more or less traditional relationships" (Strassoldo, oral communication) and the danger "that the Friulians might lose their souls."

Also in our study we attempted to tackle this phenomenon in our questionnaire and in many personal interviews. In these questions the predisaster situation was the reference point. We took the frequency of agreement with our statements as a possibility of differentiation between variables of little agreement ("less family unity," "more quarreling") and higher agreement ("Friuli less Friulian—more Italian," "everything has changed"). The stronger the agreement, the stronger the perceived change (Fig. V.17).

Seven given statements could be answered by "true," "partly true," "not true," or "don't know." Answers were weighted with factors 1, 2, and 3. Agreement got the highest stress, excluding 'don't know."

The category of "don't know" was used by only 8% on average, for the interviewees generally took a strong position. The intention was to find out whether in the judgment of the victims social life or parts of it should have changed in contrast to the predisaster situation.

Among the seven statements, highest agreement existed with "less communication" (2.41), "more anonymity" (2.31), and "worse neighborhood relationships" (2.28). Also our interviews showed a strong change for the worse: "The old social texture has broken apart. Life is as anonymous as in the big cities. Many people stopped speaking to each other because of personal quarrels during the stage of reconstruction."

But the changed settlement pattern is also responsible for this development: "Most of the houses were arranged around an inner court before, which was in joint use and where many social contacts and encounters were possible. Today these courts have disappeared or have been replaced by small lots, divided by walls" (family in Osoppo). The resident pharmacist confirms that

Fig. V.17
Opinions concerning social living

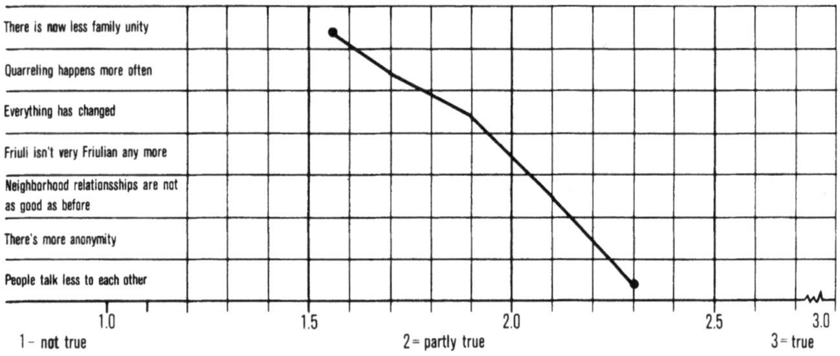

". . . the village was rebuilt, but people are not the same any longer, friendships have broken to pieces, the young ones stay apart and the older generation escapes into its memories."

Comparing the different communities, values for Venzone (2.0), Bordano (2.08), and Gemona (2.10) are relatively close together. In Bordano the deterioration of neighborhood relations is less perceived than in other communities. But the statement "Friuli isn't very Friulian anymore" gets a higher weight. This assessment might mainly be derived from the totally changed nature of the place (see Fig. V.12 and V.13), which lost all the typical characteristics of a Friulian village. It could be any suburb of a North Italian or Central European town. On the other hand the overall assessment in Osoppo is by far the most negative (2.44, see Fig. V.18). This might go back to quarreling during the initial phase of planning and to favoritism toward certain individuals. Already as early as 1980 (Cattarinussi et al., 1981, p. 110) 55.4% of interviewees in Osoppo (average: 39.5) stated that the most important things, in order to start soon with rebuilding, were "good connections." In Osoppo also, entanglement with political parties and financial interests of individuals played a strong role, excluding non-partisans from patronage.

Classified by age groups, as expected, the highest positive average assessment comes from the group 36 to 45 years old (2.0). But the differences are small, indicating that degeneration of social life is perceived by all groups in almost the same way.

Some additional remarks written on the questionnaires underline these impressions: "There is a lack of humanity and solidarity," "there is more dissatisfaction, envy and indifference," "selfishness increased, and everybody is afraid of his own shadow" (Venzone). "People are haughty and egoism has increased," "I feel isolated and very uneasy," "there is envy between the families," "you don't meet the people you used to," "neighbors are alienated," "life is lacking" (Gemona). "There's too much concrete in my place," "old structures and traditions nearly disappeared, there is no more intimacy, and isolation is growing." "Everyone is for himself only," "I had to leave my old place and cannot readjust" (Osoppo). "The earthquake taught us many things—we now know each other to the core, experienced unjustice and hatred and the loss of the unity that stood at the beginning. We saw how poor people got rich and rich people poor, because they tried to stay honest and did not know how to draw profit from the moment of disorder as so many others did. After 10 years in the desert, dispersed here, there, and every-where, we are together again, but like strangers, full of distrust and selfishness, one against the other" (Venzone).

Fig. V.18

Opinions concerning social living by commune

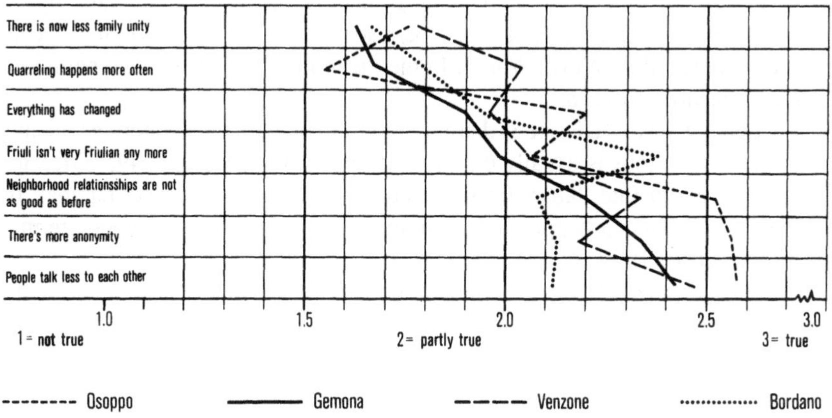

VI
Changes in the Regional Structure During Reconstruction

The paradox that destruction by catastrophe is followed frequently by an astonishing take-off in economy is not restricted to the West German "economic miracle" after World War II. In the meantime it has become a commonplace of hazard research.

To quote Cuny: "Disasters disrupt rather than destroy economies In any case, the loss of jobs is usually only temporary" (Cuny, 1983, p. 49). Friesema et al. even claim that "apart from the deaths and injuries, a disaster may be the best thing that can happen to a community, particularly in economic terms" (Friesema et al. 1979, p. 14).

Such an improvement of the economic situation at least takes place if the following factors exist:

- a predisaster high level of economy
- abundant resources for aid
- no process uprooting the population. (Klintenberg, 1979, p. 61).

The hazard-induced take-off as claimed for some cases and the neutral influence in other disaster areas are, on the other hand, opposed by a strong decline. The Val Belice and the Basilicata (Irpinia) earthquakes of 1964 and 1980 seem to fall into this category. Viewed against this background a detailed analysis of the economic development in Friuli will be given in a later part of this chapter. Here we start out with a more or less global overview, comparing the situation before and after the disaster.

1. Development of Demographic Relations

An essential object of reconstruction laws was to prevent the emptying of mountain villages and other remote areas. Therefore reconstruction subsidies could be used only in places where the applicants at the moment of the disaster had their permanent residence and had suffered material loss. On the whole this object seems to have been obtained. Statistics show no regional change in the usual long-term trends of emigration. This trend continued within the 45 destroyed communities after 1976.

The loss of 1100 persons in the few most destroyed communities was caused by the earthquake itself. In the following period there are no more losses, but the population figures stagnate at the level of 1976. Neither in the communities nor in their sections (Fig. VI.1) do the losses in population between the census of 1971 and 1981 differ between the disaster area and the adjoining districts. But it is clear also that the trend toward moving from the mountains to the area around Udine leads to an especially strong loss in the smaller, remote sections in the boundary areas of Friuli. The centers of the mountain area, however, show favorable tendencies.

Therefore, we must reject the catalytic effect of a disaster at least for the spatial variation of demographic development. But this attained objective of stabilizing demographic development may have counterproductive effects, too. The fixation of subsidies to the place of loss led to the fact that expensive new building stock was also created in places where long-existing overaging in foreseeable time will lead to the extinction of the resident population. Partly it is merely a "week-end population" who will use the grandparents' new houses in the heads of valleys for leisure time, whereas the same building stock constructed down in the valley would have provided a sufficient or even luxurious permanent residence for commuters to Udine (Photos 15-18).

2. The Socioeconomic Development

If we compare the rate of unemployment in Friuli and in the whole of Italy over an 18–year period (1968–1985), we find a parallel development (Fig. VI.2) with a quota of 1% on average lower in Friuli until the disaster. This difference grows to 3% in the decisive years of reconstruction until the two curves approach each other again in the 1980s, now with a distance of 2%. So Friuli seems to be a little bit better off even now than the labour market of the whole state. This development might predominantly go back to the sector of construction, a traditional occupation for many Friulians at home and abroad.

To local building construction, public civil engineering projects were added: the freeway Villach (Austria) — Udine was the last one to be built after the cessation of all other major public projects. It was finished in the meantime. The railway (the famous Pontebbana) will get also a second track and a new line.

The curves on unemployment show that the leap in unemployment that took place in the whole of Italy could be avoided in Friuli.

Fig. VI.1

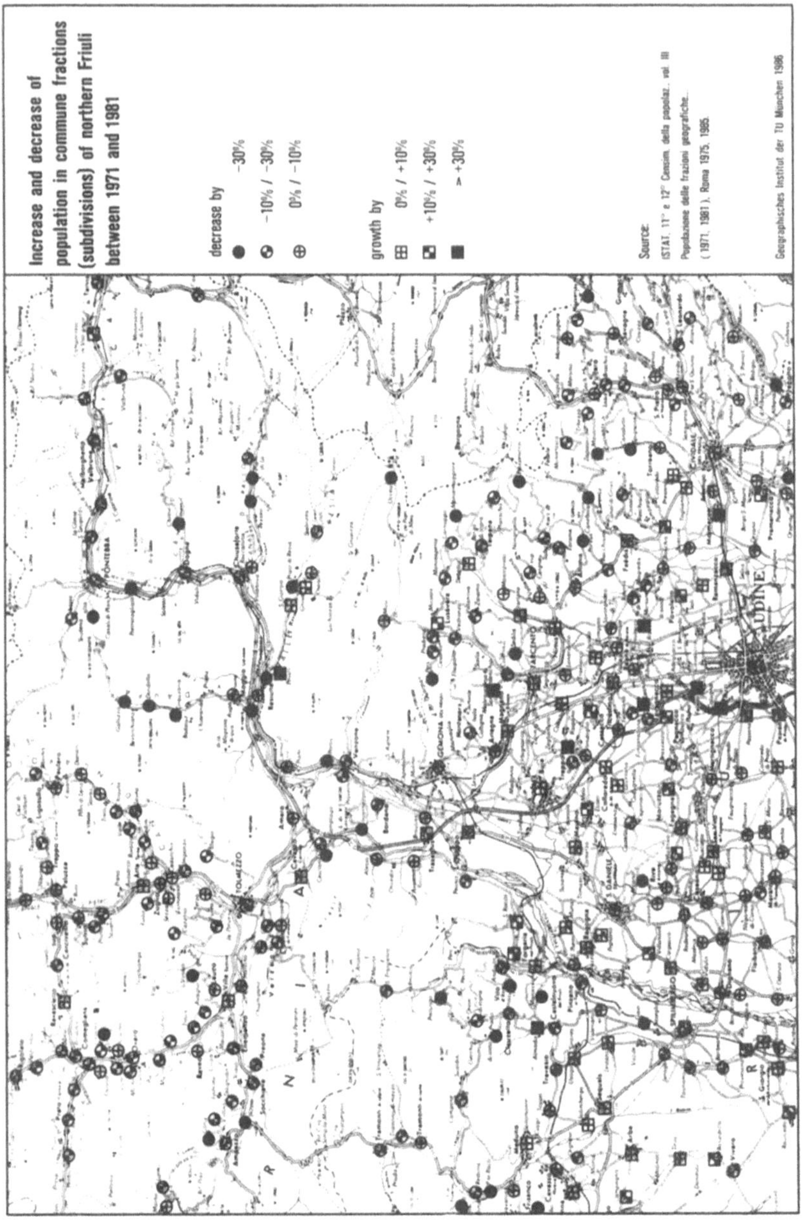
Increase and decrease of population in commune fractions (subdivisions) of northern Friuli between 1971 and 1981
decrease by
-30%
-10% / -30%
0% / -10%
growth by
0% / +10%
+10% / +30%
> +30%
Source:
ISTAT, 11° e 12° Censim. della popolaz., vol. III
Popolazione delle frazioni geografiche
(1971, 1981), Roma 1975, 1985
Geographisches Institut der TU München 1986
PONTEBBA
TOLMEZZO
UDINE

Photo 15:

Stages of an 'Economic miracle': The Pers fraction of Lusevera community in 1976 after the earthquake. The overgrown meadows in the head of the valley reveal the decline of farming.

Photo 16: In 1984 the debris has been cleared away. Only part of a cellar is left.

Photo 17: A wall in Pers whose inscription in Friulian expresses resignation: "Friul poc cognosut e Pers dismenteat" = Friuli hardly known and Pers forgotten.

Photo 18: In 1988 four new buildings surround the remnant cellar of photo 16: even in the upper valley houses are being built again - as vacation homes.

Fig. VI.2
Unemployment levels in Italy generally and in Friuli (1968 - 1985)

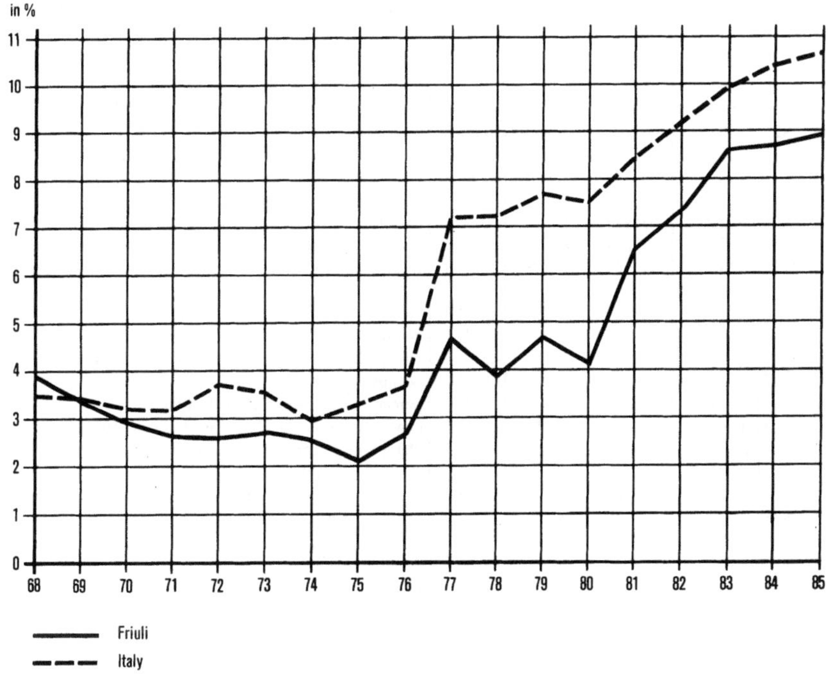

Source: Friuli-Venezia Giulia, compendio Statistico, 1969-1986

Table VI.1 Eigenvalues and proportion of explained variance of factor analysis

Factor	Eigenvalue	PCT of VAR	CUM PCT	
1	7.72638	29.7	29.7	factor "occupational structure"
2	4.65652	17.9	47.6	factor "demographic structure"
3	2.51733	9.7	57.3	factor "spatial differences of structure"
4	2.15279	8.3	65.6	factor "reconstruction"
5	1.72772	6.6	72.2	
6	1.21032	4.7	76.9	

Table VI.2 Factor loadings of variables used in cluster analysis

Var.	Factor 1	Factor 2	Factor 3	Factor 4	Variables
N1	.00880	.60765	.28620	.40828	change in percentage aged 20-44, 1971/81
N2	-.23885	-.67201	-.30018	-.11572	change in percentage over 60, 1971/81
N3	.19846	.81349	.28240	-.16632	change in increase or decrease of population, 1971/81
N4	.06883	.75148	.13954	-.20987	change in percentage with advanced schooling, 1971/81
N5	-.78159	-.35928	-.19824	-.02225	change in percentage economically active, 1971/81
N6	.07364	.23307	.89187	-.04262	change in percentage absent with workplace, 1971/81
N7	-.15730	.49113	-.18808	.09908	change in percentage commuting to outside the area, 1971/81
N9	.95177	.02367	-.09771	-.12364	change in percentage working in agriculture, 1971/81
N10	-.81456	-.20823	.08096	.02561	change in percentage employed in industry, 1971/81
N11	.07376	.62371	16156	-.31478	change in percentage unoccupied dwellings, 1971/81
N12	.09252	-.23779	-.37749	.61870	occupants of barracks
N13	.09781	.05355	-.17544	.70170	percentage of dwellings built after 1976
N14	-.05017	.11056	.91356	-.16881	change in percentage employed abroad 1971/81
N15	.75499	.17976	.14452	-.30583	change in percentage of employed women not in agriculture
N17	.17973	.69068	.17702	-.11192	present employed person who were aged 10-19 in 1971
N18	.36430	.23140	-.00468	-.75669	percentage of 20-44 years old in 1971
N19	-.30348	-.41375	-.09774	.69222	percentage over 60 in 1971
N20	.27326	-.02833	.20626	-.27726	population 1971
N21	.48113	-.17058	.27082	-.16841	percentage with advanced education 1971
N22	.71350	-.03754	-.00170	.41533	percentage economically active 1971
N23	-.11283	-.36294	-.84453	.12482	percentage absent from workplace 1971
N24	-.87509	.02789	.17334	-.06822	percentage working in agriculture 1971
N25	.42778	.36840	-.40483	.25730	percentage employed in industry 1971
N26	.02607	.48178	.03527	-.27873	percentage unoccupied dwellings 1971
N27	-.03799	-.17122	-.91134	.16574	percentage employed abroad 1971
N28	.84594	-.15807	-.15900	.23157	percentage of employed women not in agriculture
	Employment structure	Population	Spatio-structural weakness	Degree of effect	

3. Cluster Analysis of Socioeconomic and Demographic Variables

To characterize the development in the different communities of Friuli that transcend the trends of single indicators, we tried a cluster analysis of census data from 1971 and 1981. It included the data themselves and their change within that decade. The 26 variables included age structure; degree of education; number of inhabitants; economic data such as type of employment in the different sectors; commuter data; and indices of the extent of destruction, for example, percentage of prefab dwellers 1981 and the quota of apartments built after 1976, among many others.

The factor analysis immediately reveals that variables in connection with the disaster itself are of small influence on the composition of the individual factors (Table VI.2). Also, in factor 4 (affection by the disaster, reconstruction), demographic variables play a strong role.

The assignment of every community to five clusters in regard to similarities divides the provinces of Udine and Pordenone as follows (Fig. VI.3):

Cluster 1: consists of Udine, Pordenone and Codroipo, three communities with the typical factors of big towns such as youthful age groups, population growth (by 9.3%), and an oversupply of jobs (13.3% more work places than resident employees, of which 22.6% commute out in comparison to 35.9% commuting in).

Cluster 2: consists of 44 communities with an average of 7458 inhabitants and a well–balanced labor market (ratio jobs/employees 100:90), high outcommuting to cities of cluster 1, otherwise a composition of population similar to cluster 1. Only the labor market in the tertiary sector is less developed (cluster 1: 62.5%, average 43.7%).

Cluster 3: consists of 61 communities with 2990 inhabitants on average. The number of employees decreased by 3.8% which is more than the average of both provinces (2.7%). But there are only minor divergencies from both provinces' average.

Cluster 4: consists of 48 communities with 1[illegible]80 inhabitants on average. Jobs are lacking (ratio of local jobs to employees 73:100) so that 27% of employees have to commute out. These communities are overaged and show a decrease of population between 1971 and 1981 of 6.4% (average: 0.4).

Cluster 5: consists of 32 communities with 752 inhabitants on average, a number that decreased by 9.4% until 1981. Here are those communities that have the most negative trends in all other factors like age pyramid or job situation.

An influence of the earthquake on the classification of the communities cannot be found. If we compare our cluster analysis with the map of destruction (see Fig. V.1), no correlations are visible. The process of societal development has brushed aside all reminiscence of the disaster.

4. Small Scale Changes in the Regional Structure

a) Changes in the Tertiary Sector

The baraccopolises were constructed outside the old city centers and this was the case also for business, which followed the buying power of the residents by building temporary shops, bars, and bistros to serve the needs of the prefab dwellers. These shops in many cases did not remain temporary arrangements,

Fig. VI.3

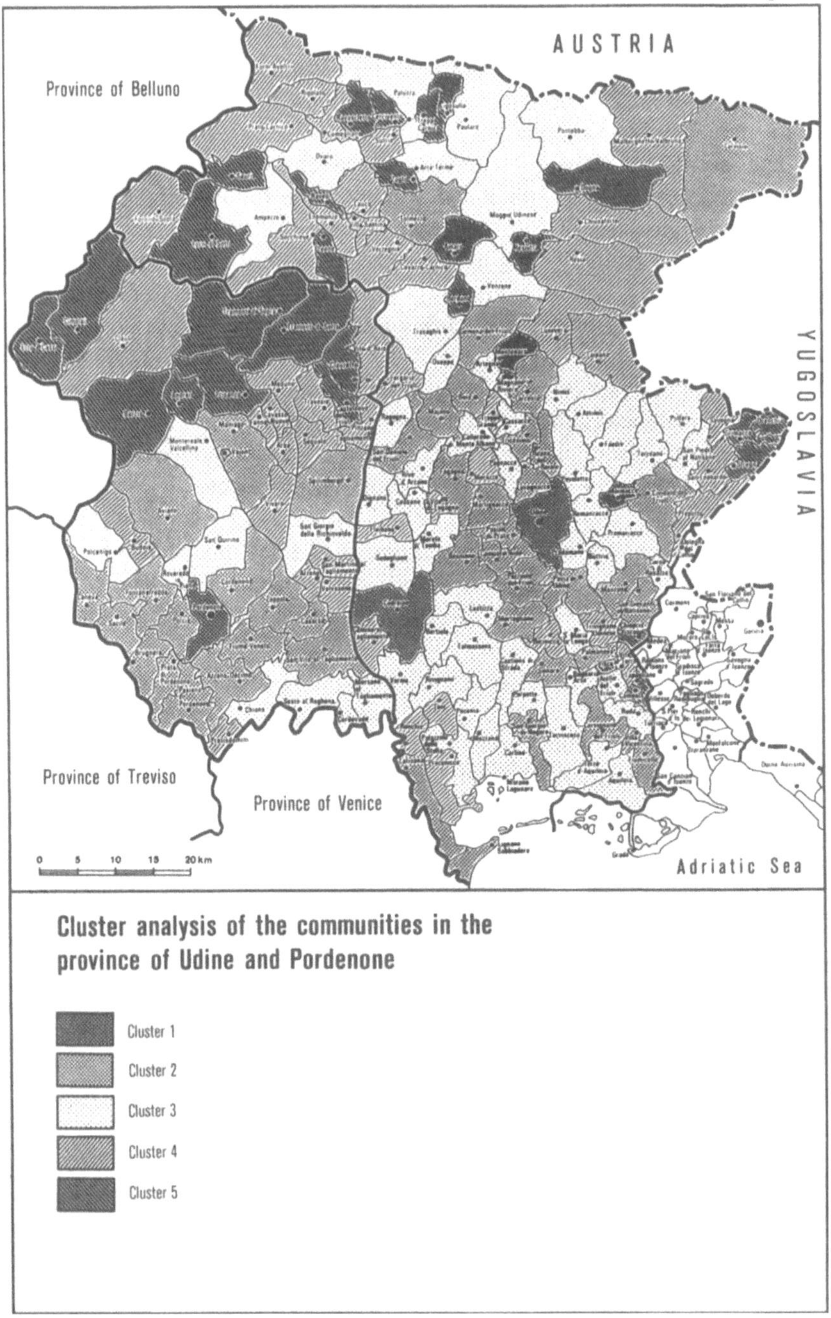
AUSTRIA
Province of Belluno
YUGOSLAVIA
Province of Treviso
Province of Venice
Adriatic Sea
0 5 10 15 20 km
Cluster analysis of the communities in the province of Udine and Pordenone
Cluster 1
Cluster 2
Cluster 3
Cluster 4
Cluster 5

because shop owners earned a lot of money. Self–supplying part–time farmers had lost their barns and stock and now had to buy food; lost utilities and articles had to be replaced even for the spartan life in the huts, and the traditional city-center shop owners in the baraccopolis offered anything now necessary. But also other shops mushroomed and became permanent, especially at well frequented locations like railway stations or bus stops. Simultaneously with reconstruction, too, the old shops in the city center were rebuilt.

Beyond reconstruction and embellishment of former shops, business owners used the subsidies to create property in the centers while still earning money at the periphery. By and by, with the expected return of the residents branch establishment and main shop could then change roles, migrating with the buying power.

The overall trend toward shopping centers along the big tourist arteries heading for the Adriatic coast and its many resorts was contrary in various cases to the philosophy of revitalizing the historic center. The evacuees got used to this "Supermercato"-type, which then commuters on their way back to the baraccopolis frequented. As a consequence of this change of attitudes, many shops in the market streets of the bigger towns are gone. Even in Gemona, a former center of trade, where the community built multistory parking decks to attract consumers uphill to the Centro Storico and its location on the alluvial fan, the empty framework of shops-to-be fills up reluctantly (photo 19).

Photo 19: Ramp of a hillside road leading into the old city center of Gemona, with parking structure for customers who will hopefully be attracted up to the alluvial fan.

Because of this Gemona now presents a triad of business locations

- at the Interstate (shopping centers, supermarkets)
- at the railway station
- at the historical center

Such a dissipation weakens the idea of planning a lively reconstructed center. Rubin refers to a similar phenomenon in Coalinga, California. "Also, as is true in many other cities, businesses are locating in shopping centers outside of the city center" (Rubin et. al., 1985, p. 271).

The same development threatens Newcastle's reconstruction of its heritage precincts in the historical CBD. While rebuilding is still debated, much of the population's buying power will be attracted by the undisturbed development farther west after the earthquake of December 28, 1989.

b) Intraregional Migration

Hazard literature frequently states that a more or less large number of individuals not only consider a change in their personal situation because of postdisaster conditions but also act to carry it through. A latent readiness for change becomes a manifest activity. Migration is the most decisive of all these considerations.

"The earthquake offered an opportunity for some people to move and they took advantage of it . . . the city offered new employment opportunities due to the earthquake reconstruction . . . the migration was produced by economic opportunity created by the earthquake and not by economic loss" (Belcher and Bates, 1983, p. 126).

Decisions that without the disaster would have never or, if at all, much later been made, or at least would have been more spaced in time, now concentrate within a short period. Belcher and Bates state that "The disasters (Hurricane David, 1979 and Guatemala earthquake 1976) probably compressed this movement into a shorter time span rather than, over the long run, really enhancing the number of migrants" (Belcher and Bates, 1983, p. 127).

In Friuli no mass emigration took place because of the promotion of reconstruction by government, which tied the people to their places of residence and even brought back emigrants—at least for a while. The Friulians were true to their statement as it was given during our first enquiry in April/May 1977: only 2% then considered emigration.

Because of the subsidies and legal framework even the large group of those who wanted to "remain in the emergency building" have rebuilt in the meantime. The subsidies in a very real sense solidified the old structures.

The exception to this principle took place, with some alterations, in Friuli between 1978 and 1984 because of paragraph II of law 70/78 and paragraph 41 of law 53/84. They granted the possibility of investing the subsidy due to

Table VI.3 What initiative do you intend to take in the near future to improve your present situation?

No reply	695	10.6%
Begin reconstruction	2685	40.9%
Move to another part of Friuli	54	0.8%
Leave Friuli permanently	11	0.2%
Move away temporarily and return later	37	0.6%
Remain in the emergency building	2833	43.1%
Start reconstruction out of the emergency building	253	3.9%
Total	6568	100%

Source: Geipel, 1982, p. 62.

an applicant not only in the individual's own community where the damage originated, but on request also in another community within a limited disaster area. Postponed and latent intentions to move for instance to a village closer to the workplace suddenly became manifest through this possiblity of transfer ("trasferimento dei contributi"). Direction and strength of this transfer of title-deeds and financial means thus revealed the tendency toward latent mobility (Fig. VI.4), disclosed by the "trasferimento." Fig. VI.4 shows that these possibilities of transfer were used mainly by individuals from the mountain communities.

Striking the balance of applications to move into or out of the mountain communities, in their eastern wing adjoining Yugoslavia and in the western part toward the province of Belluno, 90% to 100% of all applications aim at out–migration. A contingent sector of communities south of the disaster area, however, attracts applications of in-migration. Sometimes they aim farther to the south than the law allows.

Study of records in selected communities revealed the exact direction of these intentions to move. Many disaster victims wanted to move (together with their subsidies) to the central place of Gemona, while most of those citizens of Gemona who wanted to migrate applied for the next place in the hierarchyof centrality: Udine (Fig. VI.5).

In the mountains, Tolmezzo, an aspiring industrial center, inspires many intentions to migrate (Fig. VI.6), as does Spilimbergo (Fig. VI.7).

Many individuals entitled to subsidies wish to leave Bordano (Fig. VI.8). The consequence is that many houses are void, as described before. Since these people come from the most active age cohorts, the demographic structure of Bordano deteriorates.

The category "emigration with transfer of subsidies" prevails in the age group of 25 to 44 years. More than 40% of all applicants come from this most active group, which in Bordano was reduced between the censuses of 1971 and 1981, while the proportion of the old people increased.

This possibility between 1978 and 1984 benefited all those communities that would have attracted immigrants anyway. But it should be kept in mind that the possibility of transferring contributions and therefore migration was very restricted. The fundamental possibility was granted when most of the applicants had already applied for their subsidies due under the laws 30/77 and 63/77.

In any case, the community administrations in areas of out-migration were reluctant to provide information about this option. Elderly people with little education and individuals living abroad were intentionally kept unaware of the legal specifications such as "trasferimento dei contributi." This instrument was against the rules of self-preservation of small communities. This strategy of self-preservation is also dominant in the attempt at maintaining agriculture in the high mountains.

Instead of the individual barns in Lusevera that had been destroyed it was intended to bring the cattle of all those part–time farmers who were willing to join their meadows for collective grazing together in a cooperative (Photo 20). The ruins of this large construction show that attempts at revitalization of a traditional agricultural structure failed. The older generation resigned, while the younger are tied to the workplaces in the foothills and plains. The mountain village is only a dormitory. The transfer of subsidies to a community with jobs would have been the better solution.

The red-tape procedures, (application, approval by the community, investigation by the region, payment by the region, letter of agreement from the community of destination), however, were discouraging for most of the applicants. If this instrument had existed from the beginning and if the region would have propagated it, subsidies could have been employed more profitably for the individual families as well as for the region. The direction of this transfer should not have been Udine itself but smaller central places closer tothe mountains. It was Udine where most applicants wanted to move (Fig. VI.9).

For our catalog of Do's and Don'ts for a better planning strategy we have just learned that a guiding principle for the settlement pattern had to be ready before laws of reconstruction were issued. The assignment of official authority to the individual community brought about an atomizing of building activities. The insight that it would have made more sense to pool subsidies at central places (not Udine, however) came too late. It was only a correction of developments already in progress, aimed against the community egoism of the opinion leaders. But only those individuals who had already loosened their links to their community and resisted the pressure of conformity could resist the slogan of "same place." Reconstruction in Friuli is therefore the decentral-

Fig. VI.4

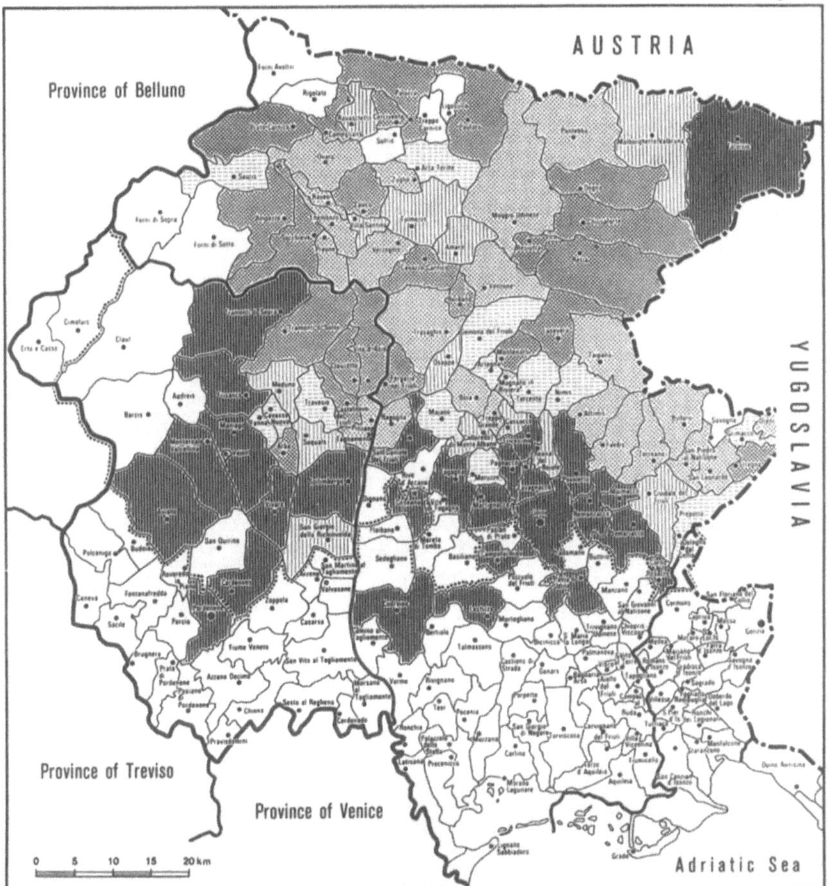

Transfer of "contributi" (Rebuilding Funds) from a commune of emigration to one of immigration, as a percentage of applications submitted

- 90 - 100% applications to move away from
- 60,1 - 89,9% applications to move away from
- 40,1 - 60,0% applications to move
- 60,1 - 89,9% applications to move to
- 90 - 100% applications to move to
- ·············· delimitation of communities, where applications were allowed

Source: Segreteria Generale Straordinaria, Udine 1986

Fig. VI.5

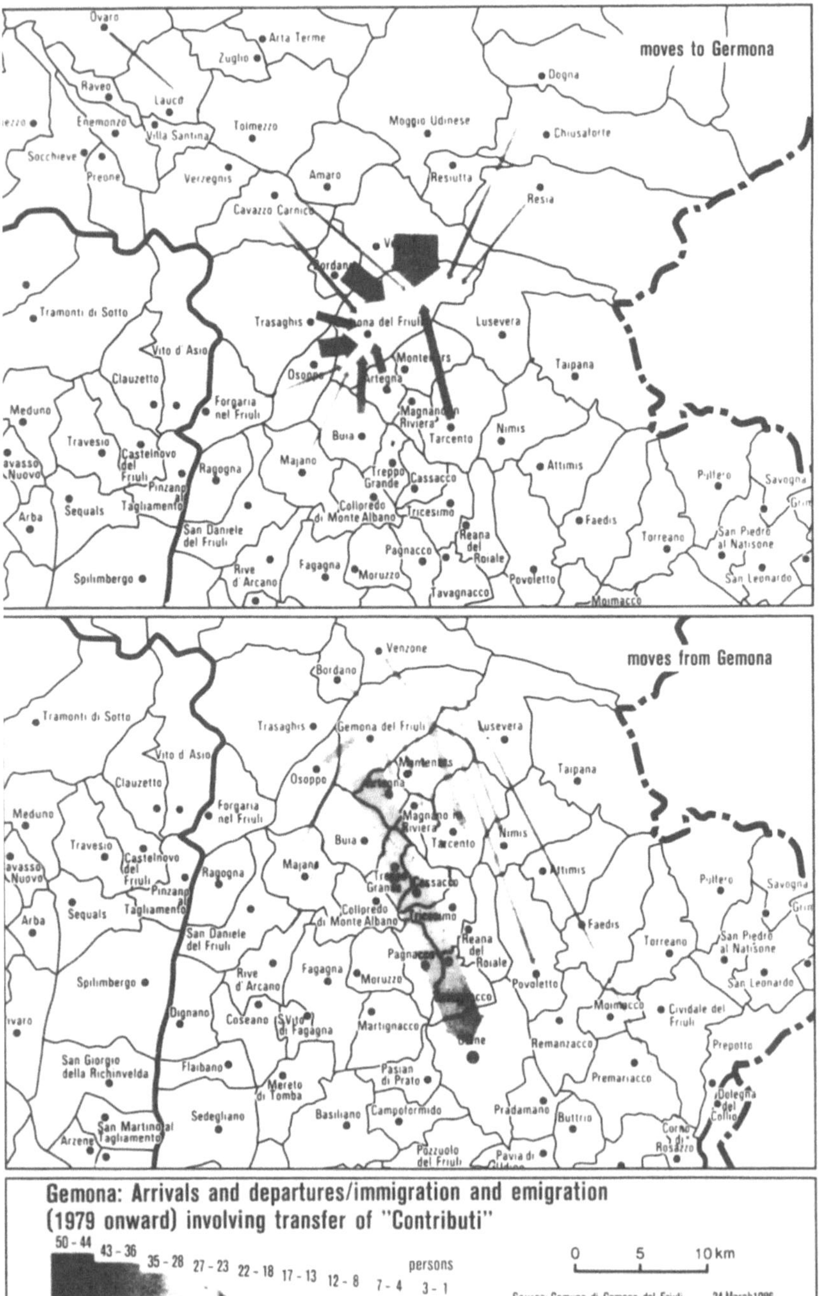

Gemona: Arrivals and departures/immigration and emigration (1979 onward) involving transfer of "Contributi"

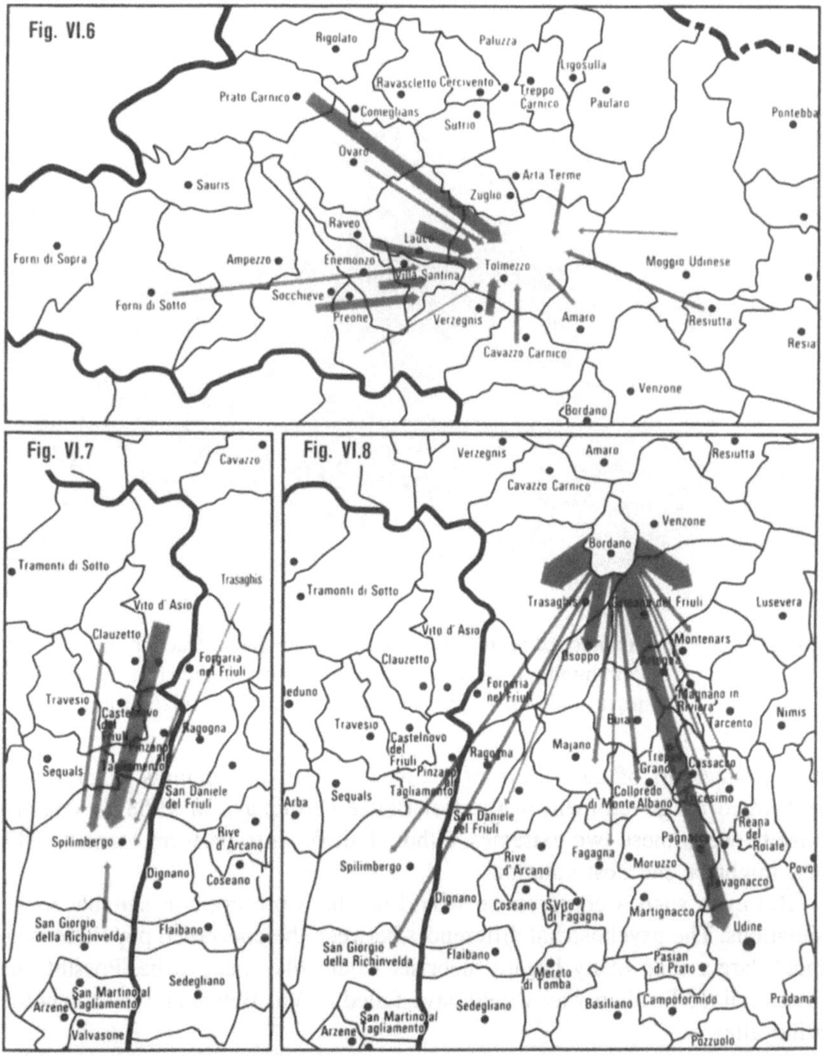

Fig. 6 persons who moved to Villa Santina

Fig. 7 persons who moved to Spilimbergo

Fig. 8 persons who moved from Bordano

persons

35 - 28 | 27 - 23 | 22 - 18 | 17 - 13 | 12 - 8 | 7 - 4 | 3 - 1

Source: own results

0 5 10 km

Photo 20: Abandoned community shed in Pradielis fraction of Lusevera community, indicating lack of confidence in the chances of survival for a traditional agrarian structure in a modernized form.

ized counterpart of the failure of planned centralized attempts as they were performed by government in the Val Belice earthquake of 1968. A compromise between these two extremes, although difficult to perform, seems viable and might be the best solution.

But even such a compromise would not be a prescription suitable for all situations. The psychosocial differences between the concerned populations are much larger than many hazard theorists would like them to be. Possibly the decentralized Friulian way would have failed in Val Belice and led to another blind alley.

5. Earthquakes, Reconstruction, and Industrial Development

a) Decline or Take-Off as Consequences of a Disaster?

In hazard research many case studies have suggested that, given a sufficient economic stage of development in the affected region and an expeditious provision of sufficient means,a disaster might even offer a chance of unforeseen development (Friesema et al., 1979; Rubin et al., 1985; Cuny, 1983,

Dudasik, 1982, Ellson et al., 1984). In the case of Friuli, testing of this hypothesis is of special interest since the Italian State had programmatically linked reconstruction with the region's industrial development. In plain words, compensation for directly *affected* plants was granted as well as low-interest loans to *all* plants of the affected area.

Measuring long-term consequences of a natural disaster is complex because the disaster-induced factor is difficult to isolate from other variables that might have influenced the economic development. Drabek comments on Friesema et al.: "They became acutely aware of how difficult it is to hold other factors constant. Thus, to claim the 'the disaster' was the sole force in operation requires the type of quasi-experimental design they employed" (Drabek, 1986, p. 237). Therefore M. Loda of our research team in her assessment of reconstruction selected among others only one economic sector: finishing industry, which is of predominant importance for the economy of Friuli, and tested it within and outside of the disaster-stricken area.

b) The Economic Structure of Friuli

The industrial structure of Friuli is based mainly on small and medium–sized businesses in the branches of mechanical industry and wood-processing and furniture. In 1981 such a plant on average had only 11.5 employees. In the course of the concentrations of the 1950s and 1960s such an atomized and small–scale industrial structure was supposed to be quite negative as seen on the background of the expanding economic centers in Northern Italy (Grandinetti and Grandinetti, 1979; Saraceno, 1981).

During the 1970s, however, when the trend of economic concentration lost its drive and in turn the high flexibility of small business gained importance, the disadvantages of a fragmented industrial structure changed into a factor of regional development. In Friuli the number of businesses increased above average and they succeeded in improving their working conditions. Emigration stopped (Valussi, 1972, Meneghel, 1983).

The earthquakes of 1976 caught Friuli right in the middle of this process of regional development, and apprehension was high that disaster might have quenched this just initiated industrial growth for good. Three hundred ninety two manufacturing plants and 2633 handicraft businesses suffered damages amounting to nearly $353 million (in 1988 values).

In turn, the generous and expeditious provision of considerable financial means gave occasion to hope. Also the mentality of the workforce who put reconstruction of workplaces above the reconstruction of homes spread optimism among the plant owners. The Friulians recognized that to prevent them from having to leave their country the right order should be "plants first, then homes, then at last churches". ("Prima le case, dopo le chiese").

Loda's survey is based on the questioning of 140 businesses (with the exception of the building industry) of Udine province with at least 10 em-

Fig. VI. 9

Applications to transfer "Contributi" to Udine
YUGOSLAVIA
Dogna
Chiusaforte
Moggio Udinese
Resiutta
Resia
Tolmezzo
Amaro
Raveo
Lauco
Ampezzo
Enemonzo
Villa Santina
Socchieve
Preone
Verzegnis
Cavazzo Carnico
Venzone
Bordano
Trasaghis
Lusevera
Taipana
Tramonti di Sopra
Tramonti di Sotto
Vito d' Asio
Clauzetto
Osoppo
Montenars
Forgaria nel Friuli
Meduno
Frisanco
Travesio
Castelnovo del Friuli
Cavasso Nuovo
Fanna
Maniago
Ragogna
Pinzano al Tagliamento
Majano
Nimis
Attimis
Faedis
Torreano
Pulfero
Savogna
Drenchia
Grimacco
San Pietro al Natisone
Stregna
San Leonardo
Cividale del Friuli
Moimacco
Prepotto
Premariacco
Dolegna del Collio
Sequals
Arba
Vajont
San Daniele del Friuli
Rive d'Arcano
Fagagna
Spilimbergo
Dignano
Coseano
S.Vito di Fagagna
Martignacco
Udine
Pasian di Prato
Campoformido
Pradamano
Buttrio
Vivaro
43 - 36
35 - 28
27 - 23
22 - 18
17 - 13
12 - 8
7 - 4
3 - 1
people involved
0
5
10 km

ployees (last census 1981). This random sample comprised about 12% of the total.

The questionnaires asked the entrepreneurs for the structural data of the plants (turnover, number of employees, etc.) and for the evaluation of the reconstruction process. These directly investigated data were then correlated with the official figures of the "Direzione Regionale Industria" for every plant in the sample on compensations, low-interest loans, and other allowances. This method provided the exceptional opportunity to combine on an individual case-by-case basis "hard facts" (officially alloted subsidies) and "soft facts" (ascertained evaluations by the interviewed entrepreneurs).
Our survey aimed at answering the following questions:

1. Did the reconstruction in Friuli influence in a positive way the development of the local finishing industry?
2. Was this influence homogeneous for all plants of this sector or were there differences as to type and size of the business?
3. Did reconstruction modify the differences between different regions of the affected areas and their internal industrial development?

c) Effects of Compensation and Loans on the Businesses

The assumption that reconstruction might have had a beneficial influence on the development of the plants has been widely agreed on by the entrepreneurs' estimates. In their opinion this influence reached its peak in 1980, four years after the quake, and decrease gradually during the following years (Fig. VI.10).

The development of investments of the businesses interviewed also shows a similar curve: without inflation, growth, at the time of the clearing of damages after the disaster, is of course highest just after the earthquake but remains still one third above the original level (with the exception of 1982/1983, a generally poor year for the national market conditions) (Fig. VI.11).

A comparison of the tendency to invest between plants *within* the disaster area and of those *outside* confirms the influence of reconstruction on the growth of the investment rate. Within the 10 years after the earthquake the average amount of investment within was double the value of plants outside of the disaster area (Table VI.4).

How can we explain the higher tendency to invest within the area affected? For the plants within the disaster area the funds for industrial reconstruction, the compensation of damages, and the low-interest loans for industrial development together were the decisive incentives for investment.

To avoid interferences between the variable of indemnification and low-interest loans we excluded the directly affected plants from our analysis. It then shows that the investments not only grew in the damaged plants but also in those that were not affected directly, although they were receiving low-interest loans.

The growth of investments caused by loans in plants that were themselves not affected was so strong that the relative benefit of compensation in favor of damaged plants was equal to the investment rate.

Investments consisted mainly in the extension of the area of the factory (Table VI.6).

Fig. VI.10
Average Evaluation of the Entrepreneurs of the Consequences of Reconstruction (n = 107)

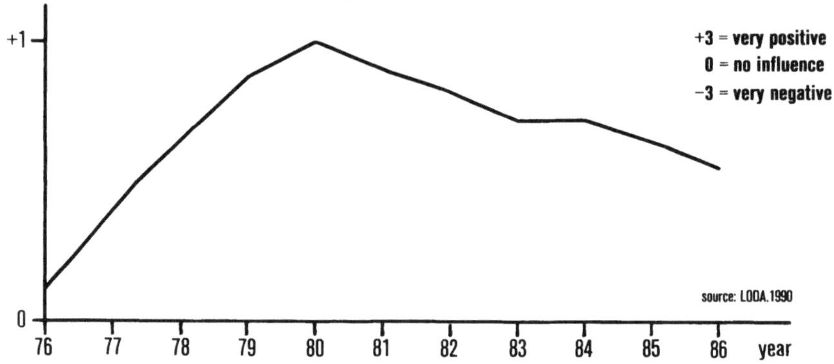

Fig. VI.11
Average Growth of Investments of the interviewed plants (prices as of 1985, 1974/75 = 100)

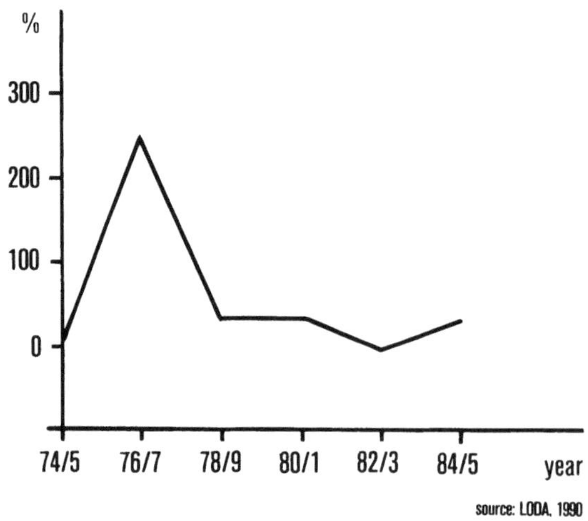

Table VI.4 Average Annual Investment Rate, by Area

	Percent of turnover
1974/75 to 1978/79	
Within the area affected	6.4
Outside of the area affected	2.3
1974/75 to 1984/85	
Within the area affected	4.3
Outside of the area affected	2.0

Table VI.5 Average Annual Investment Rate, by Access to Low-Interest-Loans (lire as of 1985, factories only)

	Average in million
1974/75 to 1978/79	
Factories with loans	343.2
Factories without loans	62.0
1974/75 to 1984/85	
Factories with loans	324.9
Factories without loans	152.7

Source: Loda, 1990

Evidently compensations were mostly used for the purchase of new ground.The built-over surface of the interviewed plants grew from 3290 m^2 in 1974/1975 to 4460 m^2 in 1984/1985, or about one third.

Comparison of plants with special low-interest loans and without them shows that the extension of factory buildings goes back to this factor. Within 10 years built-over surface of plants with credits grew for 3000 m^2, without such credits only for 650 m^2.

The compensation led also to *renewal and extension of the machinery*. Engines constructed before 1975 make up for more than one third of all machinery of the sample, but for plants with low-interest loans this proportion is below 25%. Comparison of plants of different sizes and branches gives evidence that the process of modernization was not confined to certain groups. All types of plants improved their machinery if they were granted low-interest loans.

Special credits also brought an increase of value of the technical equipment, in privileged plants more than three times more than in plants without low-interest loans (1496.9 million lire compared with 503.1).

If we look at the *directly* affected plants, however, we do not find invest-

Table VI.6 Areal extension of 139 plants, by damage

	Extension in percent Yes	No
Damaged plants	75.9	24.1
Undamaged plants	44.5	55.5

Source: Loda, 1990

ments above average into the renewal and extension of machinery. It is also evident that economically unstable smaller plants cannot draw as many advantages from special loans than bigger factories.

Negative effects of natural disasters therefore make themselves much more conspicuous in economically unstable small plants, whereas large-scale enterprises draw greater advantage out of the preferential treatments of the reconstruction process, which confirms the findings of Foster (1980, p. 252) and of Friesema et al. (1979, p. 79). However, if we study the overall performance of the plants under consideration, the effects of reconstruction are less positive than the investments just mentioned. They are even partly negative.

Loda's survey for instance discovered no real innovations in the plants under consideration, and if so, more among the damaged ones (Table VI.7).

If we consider the technological efficiency of the plants, the effects of reconstruction are even negative. The share of the costs of production from turnover rises from 44% in 1974/1975 to 47.3% in 1978/1979 and even 47.7% in 1984/1985. While compensations for the damaged plants led to a growth of efficiency, low-interest loans in the long run had a reverse effect.

As to turnover, the same effects as described by Foster and Friesema et al. showed up (Table VI.8).

The damaged bigger enterprises nearly doubled their turnover against an increase of only 10% of the undamaged ones. In turn, the damaged small enterprises lost out to the undamaged ones.

Table VI.7 Percentage of Innovation, by damage

	No	Some	Quite a few	Many innovations
Damaged plants	15.8	31.6	15.8	36.8
Undamaged palnts	22.1	30.8	31.7	15.4

While the influence of reconstruction in regard to real estate property, building stock, and machinery is clearly positive, the economic performance of the enterprises under survey is much more uncertain. Compensations brought about an enlargement and in some respects modernization of the middle–sized and large–scale enterprises. This did not happen with the smaller plants, thus empirically confirming findings of hazard research.

Compensations indeed brought about an early recommencement of production and were in favor of an extension and improvement of internal structures. Low-interest loans, however, proved to be almost counterproductive under economical aspects.

d) Regional Effects of the Financial Input

Ten years of reconstruction provided the affected enterprises with an increase of real estate capital, but not with a significant economic growth or a reasonable modernization of the productive factors. The tendency to an extension of area and built-over space is three times stronger in the disaster area than outside. An estimation of 1.5 million m^2 growth of the built-over area calculated from our sample would assign 94% to the disaster area and only 6% to the rest of the province. But such an extension of real estate capital instead of investment in new products, new production methods, and innovations of marketing indicate a suboptimal use of resources. When everyone builds, the entrepreneurs evidently were induced to do the same. They also were caught

Table VI.8 Percentage of Growth of Turnover between 1974/1975 and 1984/1985, by Size and Damage

<u>Middle-sized to large-scale enterprises</u>	
Damaged	186.8
Undamaged	110.8
<u>Small enterprises</u>	
Damaged	136.5
Undamaged	246.0

in the prevailing mentality of "more and bigger and more beautiful than before." The loans were handled through local banks whose manpower and know-how were not accustomed to the huge influx of money in such a short period and were trained to consult their traditional customers.

Also, the regional disparities did not disappear during the process of reconstruction. Reconstruction laws indeed had provided special incentives in favor of the mountain areas. But only 5% of the industrial subsidies finally made their way to the north. The highest demand for subsidies came from the well-developed hillside and plains communities that had suffered only minor damage from the earthquake.

The last industrial census before the disaster (1971) showed 12.8% of all enterprises enterprises of the finishing industry with 11.5% of jobs in the mountain zone. The figures of 1981 with 11.2% respectively, 10.3% showed that locational decisions of entrepreneurs had been in favor of the south (Fig. VI.12).

If we consider the average proportion between the subsidies granted to the loans given, the mountain areas with 1:1.7 were better off than the hill zone with 1:1.9 and the plains around Udine and Cividale with 1:2.2. The Italian government because of this preference for the weakest area succeeded in at least stabilizing the development, which all over the Alpine region is in disfavor of the mountains compared with the plains. The disaster could have finished all industrial development in the mountains. The measures taken could not stop the general trend of a loss of attractivity. But they succeeded in securing the status quo.

Fig. VI.12

Location decisions for New Establishments, Delocations and Branch Establishments

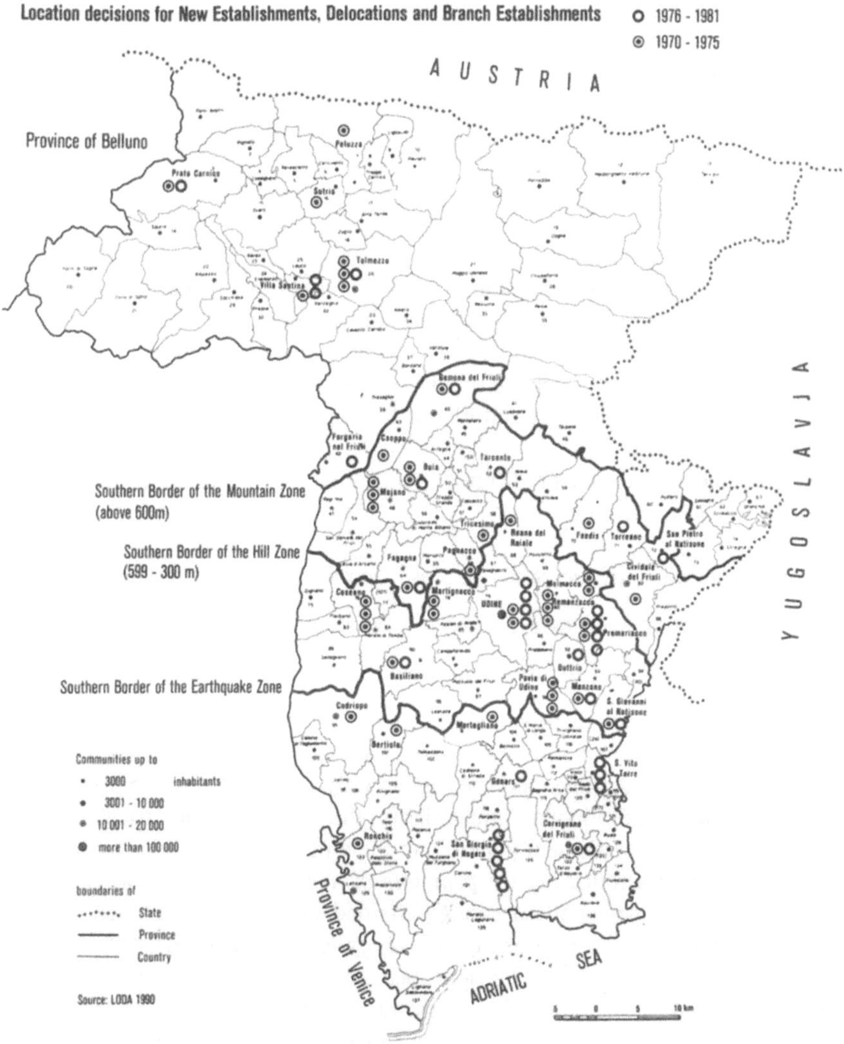

VII
Conclusions

Reconstruction following a natural catastrophe always includes the necessity and chance to change structures—either at the victims' cost or subsidized by society. The need to act finally leads to changed spatial structures. Character and scale of these changes depend on the aims chosen and on given means and resources. Against the background of our findings in Friuli we shall evaluate aims and instruments.

Despite the abundance of different aims they can be divided into three categories. These are either

- the restoration of the status quo as completely and rapidly as possible
- reconstruction seen as a chance to realize structural improvements
- or no defined aim for the time after the disaster.

In reality, of course, there is a simultaneous co-existence of more aims, of which only a few are explicitly formulated and which also can contradict each other. Mostly contradictions become visible only on the level of implementation. They can co-exist having equal rights: the aim of spatial structuring to restore the previous settlement pattern and the aim of social politics to distribute public resources following the *principle of need.* It is clear, however, that the intended settlement pattern can be attained only with the choice of the social option to compensate for *individual losses*, not for needs. Only if the individual owner gets subsidies proportional to his losses is he capable of restoring the predisaster status. The distribution of subsidies following need leaves ruins wherever the owner is not capable of restoring by his own means a multistory building or an old palace. On the other hand, this strategy might enable many persons in want of a home to acquire a lot and build a house.

In principle, available instruments are unlimited. They include the passing of laws for compensation and its criteria as well as compulsory measures for reformulation of building codes, forms of propaganda, and manipulated information as well as governmental steering of all activities (or abstention from doing so).

Between all of these different aims and instruments, a vast range of combinations can be imagined. We will confine ourselves, however, to possible forms of reconstruction following the three examples outlined above. They mirror in an idealized typology the different course of the various stages. The background might be the events in Friuli, Naples, Mexico City, or Armenia. But this is a scenario type of description of three alternatives that does not lead to strict "technical instructions" drawn from confirmed hypotheses, but rather to a conclusion derived from empirical findings as well as from observations and discourses that are not quantifiable.

1. Reconstruction Alternatives — Three Scenarios

Alternative 1

The policies and aims of reconstruction are not defined. There is no strict application of instruments, for example, incoming loans from foreign countries are diverted by top government to more profitable sectors of the economy than reconstruction in the disaster area.

Reconstruction in this case is then characterized as follows:

Short–Term Stage (1 year)

Relief supplies are consumed by "first aid actions" and, in part, seep away. The first wave of solidarity within the afflicted area is not used for joint efforts. There is neither planning nor steering because local authorities are completely overcharged and have to rely on aid from abroad. There is no political decision making, laws are enacted reluctantly or not at all, and neither master plans nor directions are given. The local level does not know how to act or, even worse, gets changing or contradictory directions. Subsidies are promised but seep away somewhere, or they are bestowed in accordance with financial and/or political power structures. Relief supplies are now only trickling in, because no positive reinforcement takes place among the helpers, since no positive resonance can be observed among the victims. There is no 'positive news' for the media. Apathy gains ground. Free areas or buildings are seized by squatters, especially in the vicinity of big cities or workplaces. Reconstruction is declared finished already at the stage of makeshift housing. Central government and foreign relief organizations withdraw from the disaster area. The catastrophe disappears from the headlines of the world press. Even the debris will not be cleared away any longer.

Second Stage (2–5 years)

Social differences grow visibly because governments let things take their course and free enterprise enables those who have access to decision makers and connections with the local establishment to draw off funds of the central or regional government for their private aims. Local decision makers are mostly dependent on the influential persons who have these "good connections." Parts of the funds from the beginning are detoured from the disaster area. Luxurious villas are juxtaposed to tents and huts of corrugated iron. The middle classes are weakened, and the gap between rich and poor widens. Funds diverted to the rich are used up for the import of luxury goods from abroad so there is no demand for local goods because the losers cannot and the winners will not buy them.

There is no economic development. Only small crafts oriented to the simple techniques of family firms, which cannot supply additional jobs, may come into existence. There is no basis of trust between people and government, so that administration is not motivated to invest funds in such an "unreliable" province. All payments aim at the procurement of loyalty. There is no self–confidence, no pride about ones own performance. Apathy and depression, and escapism to drugs and alcohol prevail. People wait for help from outside.

Long-Term Development (6–10 Years)

In comparison to predisaster conditions no development has taken place, while surrounding regions have made progress in their development. Regional disparities have grown. An economically obsolete, backward structure rules the

Alternative 2

Tab. VII.1 "Normality" as a goal of reconstruction, and its realization

Primary goal	Return to normality (status quo ante)	
Secondary goals	Settlement structure: Restoration at the old locations	Social structure: Compensation for individual losses
Form of organization	Decentralized planning on the community level	Reconstruction as responsibility of owner
Implementation	Fostering of repairs and new construction in the old location	Disbursement of funds to private individuals according to building accomplished

disaster area, a satellite economy dependent on more developed regions. Natural disaster has turned into social disaster. Now and then there are riots among the squatters with police, national guard, or the military moving in. Encroachment of the military, such as ejection from squatter settlements, and breaking up of cooperatives as "nuclei of subversion" are among the defensive instruments of the ruling powers and their interests, and arouse emotions and counter-reactions among the lower stratum of the population. Frequently the-disaster has been used to freely elected community administrations or ethnic minorities in the area to the surveillance of emergency commissioners, the police, or the military. Regulatory factors such as media, trade unions, citizens' organizations, or local administration are deprived of influence. Central government impedes the information flow from the disaster area to foreign countries because all news is bad news. Foreign aid (theoretically still possible) stops for good.

Short–Term Stage (1 year)

An important instrument is the delegation of reconstruction to the communities and disbursement of the funds through their authority. There is no possibility for individuals to move together with their subsidies to other locations. Instead, everybody has to rebuild at his former place. Either losses are compensated or fixed sums are paid according to family size or to other criteria. At the beginning there is high solidarity in the community ("we're in the same boat"). Community administration is the institutionally given decision maker and organizer of all actions. There is hardly any red tape. Actions are facilitated because nobody wants to climb out of the bandwagon and openly strive for his individual goal. If something is not clear, the citizens know whom to approach and where to go, for instance to get applications approved or to get the necessary forms. There is a personal structure within bureaucracy. The higher the solidarity ("our village will be reborn") the more quickly and more effective are the reactions. Coherence is strong. Since everybody, and everyone's own community, is responsible for anything that happens, everybody has the feeling that reconstruction is a vehicle to establish a better community life in the future.

The slogan of the community and its citizens is to return as quickly as possible to their own homes which means a small amount of planning and guidance, few bureaucratic procedures, and as many private rebuilding efforts and repairs as possible. Repairs are faster than new construction. The more the community confines itself to pure servicing, the more content are those who will and can organize rebuilding by themselves. Personal connections improve and speed up actions during the first stage, but they cast a first shadow on the second stage where decisions of principle have to be made. Field reallotment, reorganization of title-deeds, and unifications take time. Differences of opinion all to easily become personalized. The dependence of communication on personal connections may soon have a negative outcome.

Second Stage (2-5 years)

By and by the community grows up again on its former layout with slight accommodation to changing needs. Increased car ownership, for instance, requires garages and parking lots. The citizens received their subsidies from the State, but they also invest their own savings and a lot of practical work into rebuilding. There is an all-embracing climate of workaholism because everybody is engaged in do-it-yourself every weekend in house or garden. The further reconstruction proceeds, the more luxurious and wasteful buildings come into existence. Some families can afford but the facade. There is a silent competition for the most beautiful and largest house. The communities' task gets more and more complicated: during the first stage with its high degree of solidarity a resolute guidance was not necessary. A slight direction or "canalization" of citizens' activities seemed sufficient. But now controversies must be settled, compromises made, individual interests rejected, and common interests enforced. Just that is extremely difficult because of the close personal connections between administration (mayor, councilmen) and the voters. The smaller the community (and the more personalized informal relations), the more difficult everything appears. There is a tendency toward nondecision, so that more and more exceptions to the rule take place. Previous exceptions to the rule become the excuses for new ones. The solidarity of the community falls to pieces. The former power relations within the community are reestablished so that "important" people are capable of extorting exceptions: illegal building, illegal expansions into the outskirts. The fight for money is also a fight for community power and for seats in the council. Connections and inside information become more and more important. Young people, who have no financial advantages in the community such as real estate, houses, or subsidies, move closer to their workplace into central places, and other groups also in the face of dwindling solidarity now dare to move elsewhere, where the jobs are. The process of demographic selection, interrupted during the years of frantic reconstruction and latent before, sets in again.

Long-Term Development (6–10 years)

Short–term developments increase and get stronger. Solidarity continues to decrease. There is less communication and people have to pay for services, where before they could trust neighborhood relations. Even within families tensions grow. You have built for the whole family, but some of its members have moved away in the meantime or return only for periodic visits (weekends, holidays). As a consequence, the residences are much too big, and those who moved ask for their share because they were counted into the allotment of subsidies. The individuals now living in their big new houses, having achieved the goal for which they have worked for many years, toiling hard every free minute and without time for leisure and reflection, now begin to meditate on the reason for all those hardships. The results are disillusionment and disappointment over ungrateful children, for whom all this rebuilding had been undertaken. Neighbors, the community, and the state are accused of not interfering with this building orgy. There is a search for new meaning, new tasks, or relaxation. Consumption of alcohol and television entertainment increase rapidly. The external orientation of the younger, more active groups grows, and out-migration takes place. On first view, the community has recovered but on second view it has broken apart in the struggle of interests. Those who repaired so that their house kept its dimensions and those who built in observation of the relatively small lots in the historic center are grudging because the nicer and more snobbish buildings were constructed in the outskirts. The newly built center seems sterile and old–fashioned, flats here are empty. There are no structural improvements in respect to jobs, shopping facilities, and so forth, because all interests were concentrated on each family's own house, not its environmental ambiance.

Reconstruction is overambitious because every adminstration hopes and plans for immigrants from neighboring communities. Optimism is a communal duty but one does not cooperate with possible competitors for growth. Future developments are not or only to a small extent taken into account. You build for, say, 1976, not for 1996. Predisaster trends, if positive, are estimated optimistically high (5% population growth per year); or if negative, interpreted as susceptible to change. So in both types of communities too many buildings get built.

Trends of regional concentration are not reversed, but only delayed for a period of 5 to 10 years and slowed down. Reconstruction, however, is only an interlude within the normal trend. Consequently in many places money is not invested with a view to future development and is therefore misplaced. Indirectly, the mobile, active taxpayers oriented toward work were punished in favor of inactive, older, and sedentary groups.

Alternative 3 (Table VII.2)

The third alternative has the goal of using the disaster as a means for creating new spatial structures. Predominant are improvements of the settlement pattern: distribution of residential areas, work places, and within the single community, changes in the city center, building of new housing projects, and the layout of traffic lines.

Whereas scenario 2 took many elements from the factual events in Friuli, and while scenario 1 reflects the results of many failures in reconstruction worldwide, we have to confess that scenario 3 is still mostly visionary, a utopia yet to be achieved. In reality, under the impact of emergency and restoration,

Table VII.2 Structural improvement as a goal of reconstruction

Primary goal	Structural improvement	
Secondary goals	Settlement structure: Decentralized concentrations	Social structures: Identical Aid to all Affected
Form of Organization	Central planning at a higher administrative level	Cooperative and state reconstruction
Implementation	Regional plan land use plan new construction	Public payments incorporation of occupants into houses

planners rarely develop the sensitivity for a far-sighted regional planning. And at the moment when there would be time to think things over, the juridical framework and the prevalent conditions will have become fixed again and resist rethinking. This was the case in Friuli. The Regional Master Plan, issued a few months before the disaster struck, was by no means a welcome basis for reconstruction planning. In fact, it was hardly used at all. But the following third scenario was derived from Friuli insofar as it was identical to some individual attempts of communities in their planning of reconstruction.

Possible instruments for the implementation of this third alternative could be:

- centralized payment of subsidies by special administrative units
- a channeled concentration of subsidies to selected areas
- the concentration of funds for infrastructure and identification of residential land in selected communities only; no "sprinkling can principle"
- preconditions: a strong leadership, good planning, good data base, and strict control. Plans and laws should be ready for use in the desk drawers of decision makers.

The problematic points of such an approach will be discussed in our next paragraph.

The central administration must be in charge of all applications, subsidies, and acts of law. This bureaucracy, however, has to be capable of acting "unbureaucratically" insofar as it must use the willingness on the part of the population to act in order to prevent its sliding into apathy. Without drive from the very beginning, reconstruction will not achieve its proper take-off (cf. Chiavola's "danger points" in Fig. IV.6 between setting up of prefabs and the start of permanent construction). Even with an optimal performance, central administration is never fast enough in the view of the victim's expectations, especially not for the active groups and their financial resources. This group will always attempt by way of community administration to bypass the central administration's control. Therefore, there will be resistance of community power against central government.

"Cruelties" like interdiction of building, enacting of seismic safety regulations, and categorization of communities into groups of different entitlement to subsidies must be decided early, while the disaster is still echoing in the minds of the victims. Since a realistic assessment initially results only in planning, reformation of laws and dealing with applications, the victims and the communities concerned will always accuse the central administration for its irresponsible pussy-footing, which people must attack in support of their own interests and so judge harshly. That goes for all communities, so that 'facts' will be created everywhere (illicit building, etc.) During this stage an exodus from the weakly structured communities does not yet take place, since common solidarity prevents it.

Second Stage

It takes about three years to set the frame and to elaborate more detailed prescriptions for reconstruction, a period of time that may create discrepancies between the realities that have taken place in the communities in the meantime, and the concepts of the regional administration concerning structural improvements.

These facts may be tolerated following the slogan of "better to have few structural improvements than none." But then all those who followed the rules and refrained from private illicit building are furious and it is doubtful whether any structural improvements can be enacted by the region at all. Or the "faits accomplis" of the illicit buildings will be eliminated by bulldozers, which arouses the angry feelings of all trespassers and malicious joy among law–abiding fellow citizens.

Administration will try to compromise by including these "faits accomplis" into their own concepts if possible. Immigration into the central places-to-be starts off a building boom. Since it is mostly the younger age groups that migrate, high investments into kindergardens, schools, and so forth are necessary. This speeding up of the normal trend will lead later on to oversized institutions.

There will be a confrontation of two groups of communities: those with emigration and those with immigration. Both will ask for money from administration. The most plausible consequence is that both groups will invest far beyond their real needs. Cooperative and public building is easier and cheaper to control and survey, so that private rebuilding should be reduced. The consequence is a retardation of building because everything is supposed to happen at the same time and as quickly as possible so that costs are much higher than expected. This leads to dissatisfaction on the part of the victims who have to wait and cannot do things by themselves.

Long-Term Development

Many individuals have to stay for a long time in prefabs. A calming down sets in when more and more people can move into their new homes. But also dissatisfaction with the residences erected in public and cooperative reconstruction sets in.

The concentration of means leads to a smaller long-term squandering of funds. The favored communities are interesting to industry and the tertiary sector. Commuting distances are shorter now and the number of commuters has decreased. Social relations in both types of communities (young, dynamic, and energetic versus old, static, and weak) have changed and in the weak communities, of course, for the worse.

The region as a whole will have little in common with the predisaster area (in contrast to scenario 2). Regional disparities between "good" and "weak" communities are high.

Structural improvements through reconstruction may have taken place under favorable conditions. It is doubtful whether the long planning stages were tolerable for the victims. The persistence of still existing or newly created facilities prohibits planning from a tabula rasa and therefore real improvements. These are possible only on a small scale and can be more realistically regarded as aid for private activities, which should not be impeded during the first decisive years (if they are not totally counter–productive to the official goals).

It is important that the planners themselves provide the necessary incentives, that is, material inducements. For instance, a higher subsidy (+25%) and developed land for building should be provided for those who wish to transfer their contributions to a more dynamic community, in exchange for property in their old one. To "create inducements" could also be done by fast public or cooperative building, thereby attracting immigrants to those communities supposed to be the future growth poles of the region.

2. Learning from Reconstruction: The Do's and Don'ts of Friuli

We may have displayed enough illustrative material by now to come back to the initial goal: to formulate our results in "rules." They have been derived from the example of Friuli, however, and we have to discuss whether these conditions may be generalized or whether the uniqueness of any disaster prevents us from doing so.

1. All subsidies for reconstruction represent governmental acts with spatial implications, in fact policies of regional development. The more conscious all actors are of these structural effects, the more efficiently undesired secondary effects of reconstruction can be avoided and desirable effects achieved.

Regional effects in Friuli range from an unbalanced economic development in the highly subsidized North of the Region Friuli–Venezia–Giulia (provinces of Udine and Pordenone) in contrast to the South (with the capital Trieste, suffering from all present problems of ports) to the misplacing of funds by excessive building in areas of out-migration or to the local changes in settlement patterns.

2. There should exist concepts and master plans, which after a disaster should render possible a new "future-proof" layout for the area. Concepts like "dispersed concentratio," "pointaxial structures," or "axes of development" should guide reconstruction initiatives from the very beginning.

Although Friuli has a long history of earthquakes and a take-off of regional development, the goals of reconstruction ("where it was and how it was before") were not harmonized to these facts. The instrument of transferring subsidies ("trasferimento contributi") could have been used much more purposfully and gainfully to reach desired goals of spatial development.

3. As a minimum of far-sighted planning seismic zoning and building regulations (earthquake proofing) have to be fixed by legislation.

In Friuli the area around Tolmezzo already before the disaster had been classified as a seismic risk zone. The extent of damage was much smaller than in areas in comparable distance to the epicenter (see Fig. II.3). Also many houses in other villages that on private initiative had been constructed with the aid of antiseismic methods could be made habitable again through mere repairs.

4. In case of disaster, evacuation of as many individuals as possible is an intelligent strategy. Conflicts and problems can be minimized if evacuation is limited to a certain time only (and this must be known from the beginning), if a concept for a subsequent emergency housing exists, and if an efficient organization is capable of enforcing all measures.

In Friuli evacuation could function well because of the nearby hotels at the Adriatic coast, which were used only seasonally by tourism. Before the start of the season all evacuees were back to their home area.

5. For the long period between restoration and the occupation of newly built houses, housing in prefabs is a convenient solution. It makes possible a rather normal life, without losing sight of the final goal. It provides the time span necessary to build houses carefully in accordance with the victims' demands and aspirations.

For the assignment of prefabs some rules should be obeyed;

a) The allotment of prefabs of different quality should be avoided at least within one village because it might lead to social tension. Prefabs should be equivalent.

b) Housing in closed prefab towns (baraccopolis) is cheaper but brings more tensions. A dispersed accommodation (on the site of the destroyed house) is partly more positive for the inhabitants, but more expensive and inefficient.

c) The baraccopolis may be used as a tool to concentrate the population in central places or in main residential areas.

d) From the very beginning a concept for evacuation must be ready to set limits to the use of the prefabs.

e) Empty prefabs should be demolished at once to exclude an undesirable use by squatters.

f) Old and poor people, incapable of developing enough of their own initiatives to leave the emergency housing, must be aided by social measures (from "meals on wheels" to the timely construction of old–age homes or special flats in the normal apartment houses).

g) Young families coming without title deeds and therefore no compensation rights, should also be prevented by provision of public housing from becoming squatters.

6. Subsidies for private building should be tailored to promote or enforce the evacuation of the prefabs.

The method of paying subsidies in Friuli (50%, 40%, 10%) seduced individuals into staying as long as possible in the prefabs. The last 10% of the compensation was claimed very late because the rate was too low. The family could pretend to be "homeless" in order to enjoy free housing, cooking, electricity, and so forth in the prefab. Therefore we recommend the payment of the first half of the compensation when the house is ready in the rough, and the payment of the second 50% (with adaptation to inflation) if someone will leave the prefab within 6 to 12 months. If he fails to do so, the adaptation to inflation is cancelled.

7. Compensation should not be granted according to taxation value (old houses then would be undervalued), but to replacement costs. Subsidies must be adequate to activate private capital. They should vary with the willingness of people to give their own contributions, and not with financial means.

In Friuli on average half of the costs were met by subsidies. In retrospect this seems too high in terms of the condition of the present houses. But these effects seem unavoidable if the critical threshold of participation and individual initiative is to be crossed.

8. New buildings should not seem more profitable than repairs. Repairs are more cost efficient and also save social costs.

Where repairs took place only proven costs were reimbursed. Funds for new building in turn were granted with respect to the number of family members. This was more convenient for administration, but offered chances for manipulation. Work done by the tenants themselves was not deducted from the building costs (as it was with repairs!). The payment of subsidies calculated in square meters provided unjustified benefits that favored a surplus of space.

9. Reconstruction planning should not reflect the emergency situation alone but should also keep future demographic and social development in mind. If this is neglected, new problems arise.

In Friuli subsidizing an adequate living space brought a plurality of secondary effects for the victims. On average every family now has much more living space than before, resulting in an oversupply. But too few small apartments were built for instance for the old people and new families that came after the disaster.

10. These aspects can be reduced to the fundamental problem of whether aid should be given

- according to individual loss
- according to the principle of equality
- according to the principle of need.

The first solution fixes the social status quo ante, the second guarantees supply for all (while reducing differences), and the third leads to a redistribution in favor of the poor.

The abovementioned oversupply was generated by following the principle of equality. Despite the problems described and the complaints of those who might have profited more from the principle of aid given according to the individual loss, in Friuli the second principle was confirmed because of the extensive external funding.

11. Another fundamental decision is concerned with the centralized or decentralized decision making for reconstruction. A centralized reconstruction facilitates the placement of preordained objectives, but the bureaucracy that is necessary in this case alienates the people from administration. The settlement pattern may be more homogeneous but monotonous. Reconstruction enacted by the community allows for more individuality and produces a variegated settlement pattern. But communal ambition and prestigious building are fostered also, while purposeful structural change is inhibited.

In Friuli reconstruction was mostly decentralized. It prevented irreparable misplanning from spreading to neighboring villages. The different strategies, taken all in one, led to a variegated settlement pattern as quickly as the simultaneous building boom in so many communities allowed. The reverse of the coin is the competition of the communities for expansion and the affluence of prestigious buildings (city halls).

12. Desirable from the viewpoint of regional planning as central planning may be, it must be stressed that people's initiative should not be blocked until the time when recommendations, plans, and instruments finally are ready. In turn: from the beginning of reconstruction there must be ample space to pick out individual areas where growth should be permitted and local initiatives

possible. This would provide a security valve for a general dispersion of residential areas.

Personal initiative could unfold in Friuli only too well. Anyone with initiative and influence on local administration was allowed in many cases to build in the outskirts and move into his house very soon. Therefore dispersion of settlements has distorted vast areas around Udine and in the Tagliamento valley north of the town.

13. Planning for reconstruction should not unnecessarily push social change and segregation. It should be in harmony with social development, but also implemented as a steering tool, for example, to avoid the formation of ghettos for old people.

In part, reconstruction strengthened the trend toward a dual structure of social space with the young families in the outskirts and the older ones in the centers of the towns and villages. Beyond that a change in respect to building stock has taken place in the outskirts (single houses, lawns, fences, etc.), speeding up the replacement of an agrarian lifestyle, disintegration of the three-generation family, and change of leisure time activities. In the central areas, in turn, it seems that old structures were raised only to the quality standards of the 1980s and 1990s. This leads us back to the beginning of this paragraph:

14. Disaster and reconstruction are incisive events in the life of the individual and group so that regional, urbanistic, and economic considerations of social continuity should have high priority. Planning that follows only laws of a technical rationality would endanger the already injured identity.

The success of reconstruction goes back to the readiness of the people to act by themselves. This readiness is unevenly distributed: growing young families are the winners, old people feel themselves outdistanced and outcast. But the consequences of fixation and concentration of the younger ones for a whole decade on building their homes can also be imagined. Many observers and insiders of the Friulian society claim that the Friulians have "lost their souls."

3. Final Remarks

a) To the Case of Friuli

The economic miracle of earthquake-stricken Friuli has come to its end. Fifty thousand new homes have been built or repaired. Nearly $6 billion was invested by the Italian State in the disaster area. Private savings, personal work, and high debts in addition to the State's compensation have resurrected Friuli Phoenix-like from the ashes for a second time. The Friulians fare well today in respect to housing space.

But the evaluation of Friuli's ambiance today is composed of internal and external variables (intimacy, neighborhood, memories of the places of childhood, social well–being). In the hectic period of reconstruction, which was focused merely on the restoration of the building stock, this second aspect was neglected. The mass construction of monotonous stock, the longer distance to one's neighbor, the extension of the sphere of intimacy, and shopping outside of the village centers in supermarkets along the thoroughfares all symbolize a process that might have taken place in more than one generation, but here in Friuli occurred in less than a decade. In the adults' heads is still the predisaster sociopsychological situation while the real environmental conditions have changed totally in the meantime. "Environment" is now more hostile than it was before. However, this does not relate to earthquake risk but to the social environment. With greater time distance to the tremors of 1976, there is disillusion and disappointment grows that the improvement of the building stock was to the disadvantage of the social environment.

The generous handling of compensation and the much smaller "rate of dissipation" in comparison to other disasters in Italy created a building stock exceeding demand. This space is more useless the more peripherally it is located. In communities threatened by out-migration many emigrants were included in the computations in the hope that they would once return. But this is not merely an economic loss, it is also devaluing the houses of every other individual and the money and work invested.

But also in more centrally located communities building was much too abundant. In addition to the subsidies granted by the State, all individual savings and a lot of private work were invested. Now many families are heavily in debt, especially since with the end of the building boom many jobs in construction were made redundant and too much floor space is on hand. Nevertheless, all those who invested and had influence on reconstruction are content with it. So it can be postulated that reconstruction, despite side effects we cannot calculate, is viewed by the people with satisfaction. The State of Italy has more debts now, but at least in most parts of Friuli the political leadership gained loyalty.

The final evaluation of the success of reconstruction therefore depends on the weight attributed to the various dimensions. It seems that it might be positive as the region sees it.

b) To the Relation Between Hazard Studies in General and Geography in Particular

Hazard studies have to insist on the discovery of laws in order "to help improve disaster management and to guide better environmental stewardship in yet to be encountered disasters" (DeSanto, 1990). The author is therefore grateful to Tom Drabek, Robert DeSanto and David Alexander that they directed his endeavors from the description of Friuli's uniqueness (including the "five happy constellations" of its disaster) to the search of generalities.

Armenia, or Italy. But how can the detailed description of a single event that has happened somewhere in space and time be linked to this greater task? This tension between "uniqueness" and "generality" is as old as geography itself. Interestingly enough the Association of American Geographers has just revived this discussion in an Occasional Publication in the honor of Richard Hartshorne and in a celebration of the 50th anniversary of his *Nature of Geography* (Entrikin and Brunn, eds., 1989).

Isn't a study of a single event in a specific region a culmination point in the sense of Schaefer's exceptionalism and a "renewed celebration of the unique at the expense of the general" (Smith, 1989)?

Hazard theory, however, has to take into account that in testing its models in a variety of settings similarities of event, nation, applied laws, performance of disaster management, or used strategies are partly overruled by the singularities of a certain event, be it because of time, season, ethnic background, psychological conditions of the population, its capacity of bearing stress, the stamina of local administration, or available funds, to mention just a few factors. Therefore evidently both approaches are necessary and it is partly based on the intellectual training and personal background of the researcher, which side will be stressed.

The "Do's and Dont's" of the last paragraph are juxtapositions of generalities and uniquenesses and lessons to be learned from Friuli for the contexts of future disasters elsewhere.

A German geographer, raised in the humanistic tradition of Hartke's and Bobek's social geography and Hägerstrand's micropolitical studies applies some lessons learned in the context of Anglo-American hazard theory to a region in Italy of strong traditional roots, high complexity and differentiation for a period of 12 years. It was an attempt to take advantages from both sides: To become, in part, an insider of this specific region but also to perform in agreement with the demand for general conclusions.

Fig. VII. 1 Topography of South - Eastern Alps

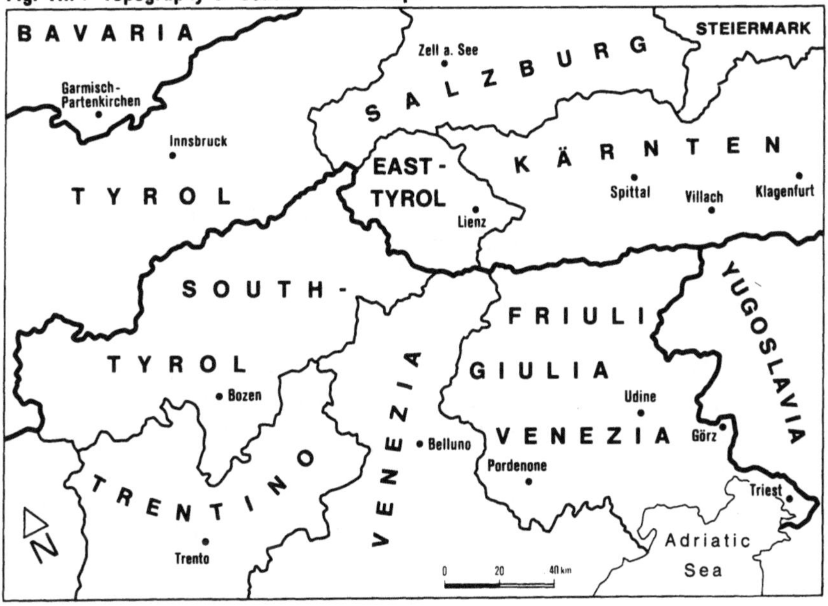

Fig. VII.1 explains the topography and the spatial relations of the colored map (Fig. VII.2 - backcover), where changes in population are shown. The emptying of the mountain areas of Northern Friuli, the disaster area, is especially strong, however, it is not limited to this part of the country alone.

Also in Southern Carinthia (Kärnten) in the northern province of Belluno and in the northern part of Trentino (in contrast to South Tyrol) population decrease prevails. Only a few central places (in Friuli Tolmezzo) react differently and attract population.

The contrast to the North-Italian plains is very distinct. In the foothill corridor from Trento via Belluno, Pordenone to Udine there is an increase of population. The counterpart of this corridor in the north are the Austrian cities of Lienz, Spittal, Villach and Klagenfurt.

Bibliography

Adamic, M.O. 1982: The effects of the 1976 earthquake in the Soca River Basin. In: Jones, B.G.,Tomazevic, M. (eds.): Social and economic aspects of earthquakes. Proceedings of the 3rd International Conference, in Bled, Yugoslavia, 29.6.-2.7.1981, Ljubljana, Ithaca/New York, pp.533-556.

Alexander, D. 1982: The Earthquake of 23 November 1980 in Campania and Basilicata, Southern Italy. International Disaster Institute.

Alexander, D. 1986: Disaster Preparedness and the 1984 Earthquakes in Central Italy. Natural Hazard Research, Working Paper #55, Boulder.

Alexander, D. 1987: The 1982 Urban Landslide Disaster at Ancona, Italy. Natural Hazard Research, Working Paper #57, Boulder.

Alexander, D. 1989: Preserving the Identity of Small Settlements during Post-Disaster Reconstruction in Italy. In: Disasters vol. 13, #3, pp. 228-236.

Allan, R. and R.L. Heathcote 1987: 1982-83 Drought in Australia. In: Glantz, M. et al. eds.: The Societal Impacts Associated with the 1982-83 Worldwide Climate Anomalies. National Center for Atmospheric Research. Boulder/Colorado.

Bagnasco, A. 1977: Tre Italie. La problematica territoriale dello sviluppo italiano. Il Mulino, Bologna.

Barbina, G. 1976: Teoria e prassi della ricostruzione.

Barbina, G. 1977: "Il Friuli centrale dopo gli eventi sismici del 1976." In: Bollettino della Società Geografica Italiana, Vol. VI, 10-12, pp.607-636.

Barbina, G. 1978: Le attività economiche nelle aree terremotate: problemi di ieri e di oggi. In: Friuli Earthquake as Civico di Storia Naturale, Triest, pp.23-37.

Barbina, G. 1983a: I terremoti del Friuli nel 1976. Bilancio di un'esperienza. In: Orizzonti Economici, n.37, pp.83-85.

Barbina, G. 1983b: Il caso del Friuli sette anni dopo gli eventi sismici del 1976. In: Atti del XXIII Congresso Geografico Italiano. Vol.III, pp.276-280; pp.290-291.

Bates, F.L. and Kilian, C.D. 1982: Changes in housing in Guatemala following the 1976 earthquake: With special reference to earthen structures and how they are perceived by disaster victims. In: Disaster, Vol. 6, Nr.2, pp.92-100.

Bates, F.L. and Kilian, C.D. 1982: The use of a crossculturally valid level of living scale for measuring the social and economic effects of earthquakes and other disasters and for measuring progress in recovery and reconstruction as illustrated by the case of the Guatemalan earthquake of 1976. In: Jones, B.G. / Tomazevic, M. (eds.): Social and economic aspects of earthquakes. Proceedings of the 3rd International Conference, in Bled, Yugoslavia, 29.6.-2.7.1981, Ljubljana, Ithaca/New York, pp.479-498.

Bates, F.L. and Peacock, W.G. 1987: Disaster and social change. In: Dynes, R.R. and De Marchi, B., Pelanda, C. (eds.): Sociology of disasters. Contribution of Sociology to Disaster Research. Milan, pp.291-330.

Baumann, J.M. and Sims, J.H. 1978: Flood insurance: some determinants of adoption. Economic Geography 54(3), p.189-196.

Belcher, J.C. and Bates, F.L. 1983: Aftermath of Natural Disasters: Coping through residential mobility. In: Disasters, Vol.7, Nr.2, p.118-128.

Bolin, R.C. 1982: Long-term family recovery from disaster. Institute of Behavioral Science, Monograph # 36 (Program on Environment and Behavior), University of Colorado, Boulder/Colorado.

Bolton, P.A., Heikkala, S.C., Greene, M.M., and May, P.J. 1986: Land use planning for earthquake hazard mitigation: A handbook for planners. Natural Hazards Research and Applications Information Center, Special Publication 14, University of Colorado, Boulder/Colorado.

Britton, N.R. 1987: Towards a reconceptualization of disaster for the enhancement of social preparedness. In: Dynes, R.R./De Marchi, B./ Pelanda,C. (eds.): Sociology of disasters. Contribution of Sociology to Disaster Research. Milan, pp.31-55.

Brown, J.M. and Gerhart P.M. 1989: Utilization of the Mortgage Finance and Insurance Industries to induce the Private Procurement of Earthquake Insurance: Possible Antitrust Implications. Natural Hazard Research Working Paper #66, Boulder/Colorado.

Brusco, S. 1982: The Emilian model: productive decentralisation and social integration. In: Cambridge Journal of Economics, Vol 6, pp.167-184.

Burlison, J. 1981: Relief work after the South Italy earthquake, November 1980. In: Disasters, Vol. 5, Nr. 4, pp.349-354.

Buttimer, I. and Pushchak, R. 1984: The status and prospects of risk assessment. In: Geoforum 15(3), pp.463-476.

Butler, J.R.G. and Doessel, D.P. 1980: Who bears the cost of natural disasters? An Australian case study. In: Disasters 4(2), pp.187-204.

Buttimer, A. 1979: 'Insiders', 'Outsiders', and the Geography of Regional Life. In: Kuklinski et al. (eds.): Regional Dynamics of Socioeconomic Change. Tampere

Cacciaguerra, S. 1986: Le esperienze e gli esiti di campo urbanistico della ricostruzione del Friuli. In: Fabbro, S. (ed.): 1976-1986 la ricostruzione del Friuli. Udine, pp.101-114, p.165.

Cattarinussi, B., Pelanda, C., and Moretti, A. (eds.) 1981: "Il disaster: Effetti di lungo termine. Indagine psicosociologica nelle aree colpite dal terremoto del Friuli." Udine, (I.S.I.G., Gorizia, Serie "Ricerche", 6).

Cattarinussi, B. and Pelanda, C. (eds.) 1981: "Disastro e azione umana. Introduzione multidisciplinare allo studio del comportamento sociale in ambienti estreni." Milano.

Cattarinussi, B. 1982: Victims, primary groups and communities after the Friuli earthquake. In: Jones, B.G./Tomazevic, M. (eds.): Social and economic aspects of earthquakes. Proceedings of the 3rd International Conference, in Bled, Yugoslavia, 29.6.-2.7.1981, Ljubljana, Ithaca/ New York, pp.519-532.

Cavazzani, A. 1982: Social and institutional impact of the 1980 earthquake in southern Italy: Problems and prospects of civil protection. In: Jones, B.G. and Tomazevic, M. (eds.): Social and economic aspects of earthquakes. Proceedings of the 3rd international conference, in Bled, Yugoslavia, 29.6.-2.7.1981, Ljubljana, Ithaca/ New York, pp.425-436.

Chamberlain, E.R. et al. 1981: The Experience of Cyclone Tracy. Canberra. Australian Govt. Publ. Serv.

Chiavola, E.: Rehabilitation strategies: The Friuli case. International Seminar "Learning from earthquakes", Perugia 11.-13.4.1985.

Cuny, F.C. 1983: Disaster and development. Oxford University Press, Inc. New York-Oxford.

Dacy, D.C. and Kunreuther H. 1969: The Economics of Natural Disasters, New York. The Free Press.

Darwin Disaster 1975: Cyclone Tracy. Report by Director-General Natural Disasters Organisation on the Darwin Relief Operations 25 December 1974 to 3 January 1975. Canberra.

Davis, I. 1977: Emergency Shelter. In: Disasters 1 (no. 1), pp. 23-40.

Davis, I. 1981: "Disasters and Settlements - Towards an Understanding of the Key Issues", in: Habitat International, Vol. 5, No. 5/6, pp. 723-740.

Davis; J. 1978: Shelter after Disaster. Headington.

DeMatteis, G. et al. 1984: Fine della marginalità alpina? Un'inchiesta presso le comunità montane del Piemonte. In: Gilbert, G. (Ed.): Regioni e comunità montane delle Alpi Occidentali: problemi economici, storici e sociali, Milano, pp. 58-83.

DeSanto, R. 1990: Letter to the author.

Dobler, R. 1980: Regionale Entwicklungschancen nach einer Katastrophe. Ein Beitrag zur Regionalplanung des Friaul. Münchener Geographische Hefte 45, Kallmünz/Regensburg.

Drabek, T.E. 1986: Human system responses to disaster. An inventory of sociological findings. New York / Berlin / Heidelberg / London / Paris / Tokyo.

D'Souza, F. 1982: Recovery following the South Italian earthquake. November 1980: Two contrasting examples. In: Disasters, Vol. 6, Nr. 2, pp. 101-109.

Dudasik, S. 1982: Unanticipated repercussions of international disaster relief. In: Disasters, Vol. 6, Nr. 1, pp. 31-37.

Dynes, R.R. 1987: Introduction. In: Dynes, R.R./De Marchi, B./Pelanda, C. (eds.): Sociology of disasters. Contribution of sociology to disaster research. Milan, pp. 13-29

Dynes, R.R., De Marchi, B. and Pelanda, C. (eds.) 1987: "Sociology of Disasters. Contributions of Sociology to Disaster Research". Milan.

Ellson, R.W., Milliman, J.W. and Roberts, R.B. 1984: Measuring the regional economic effects of earthquakes and earthquake predictions. In: Journal of Regional Science, Vol. 24. Nr. 4, pp. 559-579.

Entrikin, J.N. and S.D. Brunn (eds.) 1989: Reflection on Richard Hartshorne's "The Nature of Geography", Occasional Publications of the Association of American Geographers, Washington.

Fabbro, S. 1983: La ricostruzione del sistema insediativo a Gemona: Una ipotesi di lettura. Rassegna Tecnica del Friuli-Venezia Giulia, Estratto dal Ni. 1, Jan./Febr.

Fabbro, S. 1985a: La ricostruzione del Friuli: Un bilancio in fase di completamento dell'opera. In: Archivio di Studi Urbani e Regionali, Nr. 23, pp. 55-79.

Fabbro, S. 1985b: La Ricostruzione del Friuli. Ricerche e studi per un bilancio della ricostruzione insediativa e della riabilitazione socio-economica nell' area colpita dagli eventi sismica del 1978. Udine.

Fabbro, S. (ed.) 1986: 1976-1986: La Ricostruzione del Friuli. Realizzazioni, trasformazioni, apprendimenti, prospettive. Un approccio multidisciplinare. Udine.

Fogolini, L. 1987: La percezione della ricostruzione postsismica in Friuli. Il caso di Artegna. Quaderni dell'Instituto di Geografia. Trieste.

Foster, H.D. 1980: Disaster planning: The preservation of life and property. New York/Berlin.

Franz, S. 1979: Wiederaufbau-Maßnahmen nach dem Erdbeben von 1976 in Guatemala. Eine geographische Analyse einzelner Projekte. Inauguraldissertation im Fachbereich Geographie der Johann Wolfgang Goethe-Universität zu Frankfurt am Main, Frankfurt.

Friesema, H.P. et al. 1979: Aftermath. Communities after natural disaster. Beverly Hills, Calif.

Gazerro, M.L. 1978: Il territorio dell'anfiteatro morenico del Tagliamento: analisi dell'evoluzione socio-economica. In: Rivista Geografica Italiana, fasc.4, pp.348-363.

Geipel, R. 1977: Friaul. Sozialgeographische Aspekte einer Erdbebenkatastrophe. Münchener Geographische Hefte 40, Kallmünz/Regensburg.

Geipel, R. 1979: Friuli. Aspetti sociogeografici di una catastrofe sismica. Franco Angeli, Milano.

Geipel, R. et al. 1980: Il progetto friuli - Das Friaul-Projekt. Quaderni di "ricostruire" 1, Martin Internazionale, Udine.

Geipel, R. 1982a: Disaster and Reconstruction. The Friuli (Italy) Earthquake of 1976. George Allen & Unwin, London, Boston, Sydney.

Geipel, R. 1982b: The case of Friuli, Italy. The impact of an earthquake in a highly developed old culture. Regional identity versus economic efficiency. In: Jones, B.G./Tomazevic, M. (ed.): Social and economic aspects of earthquakes. Proceedings of the 3rd International Conference, in Bled, Yugoslavia, 29.6.-2.7.1981. Ljubljana, Ithaca/New York, pp. 499-517.

Geipel, R. 1983: Katastrophe nach der Katastrophe? Ein Vergleich der Erdbebengebiete Friaul und Süditalien. In: Geographische Rundschau 35, pp.17-26.

Geipel, R. 1986: Il significato della ricostruzione in Friuli per la "hazard theory". In: Fabbro, S. (ed.) 1986: "1976-1986", pp.115-125.

Giustolisi, F.: Modello Friuli - A 10 anni dal terremoto. In: L'Espresso, 11.5.1986, pp. 34-39.

Grandinetti, P. and Grandinetti, R. 1979: Il caso Friuli. Arretratezza o sviluppo? Udine.

Grandinetti, R. 1984: Sistema industriale e politiche regionali. Analisi e proposte per il Friuli-Venezia Giulia. Milano.

Gray, N., McKay, M., Seaman, J., Mister, R., and Davis, J. 1980: Disaster and the small dwelling, Oxford, April 1978. In: Disasters, Vol. 4, Nr. 2, pp. 140-153.

Greco, D. et al. 1981: Epidemiological surveillance of diseases following the earthquake of 23rd November 1980 in Southern Italy. In: Disasters, Vol. 5, Nr. 4, pp. 398-406.

Haas, E.J., Kates, R.W. and Bowden, M.J. 1977: Reconstruction following disaster. Cambridge, Mass. / London.

Hackelsberger, Chr.: Der Beton und seine Logik. Feuilleton der SZ, 7./ 8.12.1985, p.III.

Healey, D.T. et al. (eds.) 1985: The Economics of Bushfire: The South Australian Experience. Oxford University Press, Melbourne.

Heathcote, R.L. 1983: The Arid Lands: Their Use and Abuse. London and New York.

Heathcote, R.L. 1990 in print: Managing the Droughts? Perception of Resource Management in the Face of the Drought Hazard in Australia. Kluwer Academic Publishers, Dordrecht.

Heathcote, R.L. 1990 in print: Historical Experience of Climatic Change in Australia. Academy of the Social Sciences in Australia, Canberra.

Hewitt, K. 1982: Settlement and change in 'basal zone ecotones': An interpretation of the geography of earthquake risk. In: Jones, B.C./Tomazevic, M. (eds.): Social and economic aspects of earthquakes. Proceedings of the 3rd International Conference, in Bled, Yugoslavia, 29.6.-2.7.1981, Ljubljana, Ithaca/New York, pp. 15-41.

Hogg, S.J. 1980: Reconstruction following seismic disaster in Venzone, Friuli. In: Disasters, Vol. 4. Nr. 2, pp. 173-185.

Housner, G.W. 1989: An International Decade of Natural Disaster Reduction: 1990-2000. In: Natural Hazards, vol. 2 No.1, Kluwer Academic Publishers, Dordrecht/Boston/London.

Huffman, J. 1986: Government Liability and Disaster Mitigation: A Comparative Study. University Press of America, Lanham, Maryland.

Hultaker, Ö. 1982: Housing patterns after a landslide. In: Jones, B.G./Tomazevic, M. (eds.): Social and economic aspects of earthquakes. Proceedings of the 3rd International Conference, in Bled, Yugoslavia, 29.6.-2.7.1981, Ljubljana, Ithaca/New York, pp. 621-634.

Karakos, A., Papadimitriou, J. and Pavlides, S. 1983: A preliminary investigation of socio-economic problems following the 1978 Thessaloniki (Greece) earthquake. In: Disasters: Vol. 7, Nr. 3, pp. 210-214.

Kariel, H.G. and Kariel, P.E. 1982: Socio-cultural impacts of tourism: an example from the Austrian Alps. Geografiska Annaler #, pp.1-16.

Kates, R.W. and Pijawka, D. 1977: From rubble to monument: The pace of reconstruction. In: Haas, E.J./Kates, R.W./Bowden, M.J.: Reconstruction following disaster. Cambridge, Mass./London, pp. 1-23.

Klinteberg, R. 1979: Management of disaster victims and rehabilitation of uprooted communities. In: Disasters, Vol. 3, Nr. 1, pp. 61-70.

Kreimer, A. 1982: Housing reconstruction in the Caribbean and in Latin America.In: Jones, B.G./Tomazevic, M. (eds.): Social and economic aspects of earthquakes. Proceedings of the 3rd international conference, in Bled, Yugoslavia, 29.6.-2.7.1981, Ljubljana, Ithaca/New York, pp. 607-619.

Kunreuther, H. 1968: The case for comprehensive disaster insurance. Journal of Law and Economics 11(1), pp.133-163.

Kunreuther, H. 1974: Economic analysis of natural hazards: An ordered choice approach. In: White, G.F. (ed.): Natural hazards: Local, national, global. New York, pp. 206-214.

Ladava, A. 1982: Guidelines and procedures used to eliminate the impact of the earthquake in the Soca Valley. In: Jones, B.G./Tomazevic, M. (eds.): Social and economic aspects of earthquakes. Proceedings of the 3rd international conference, in Bled, Yugoslavia, 29.6.-2.7.1981, Ljubljana, Ithaca/New York, pp. 413-423.

Lamping, H. 1986: Angepaßte Technologien in peripheren Räumen - mit Beispielen zum Wiederaufbau nach Erdbeben. In: GS 40, April 1986, pp. 2-9.

Leivesley, S. 1980: The social consequences of Australian disasters. In: Disasters, Vol. 4, Nr. 1, pp. 30-37.

Loda, M. 1990: Erdbeben, Wiederaufbau und industrielle Entwicklung im Friaul. Eine Langzeituntersuchung. Münchener Geographische Hefte Nr. 65, Kallmünz-Regensburg.

Mader, G.G. 1982: Land use planning after earthquakes. In: Jones, B.G./ Tomazevic, M. (eds.): Social and economic aspects of earthquakes. Proceedings of the 3rd international conference, in Bled, Yugoslavia, 29.6.-2.7.1981. Ljubljana, Ithaca/New York, pp. 589-605.

Mattioni, F. 1986: "La Ricostruzione Industriale del Friuli." Udine.

Meneghel, G. 1978: La previsione dei terremoti e le economie regionali. In: Bollettino della Società Geografica Italiana, Vol. VII, pp. 141-148.

Meneghel, G. 1983: Friuli. In: Gentilischi, M.L. and Simoncelli, R. (Eds.): Rientro degli emigrati e territorio, risultati di inchieste regionali. Naples, pp. 105-122.

Milani, F. 1982: Dinamica socio-economica e articolazione territoriale nel Friuli terremotato: un tentativo di sintesi, Istituto di Geografia, Università di Padova, Padova.

Mileti, D.S. 1987: Sociological methods and disaster research. In: Dynes, R.R./ De Marchi, B./Pelanda, C. (eds.): Sociology of disasters. Contribution of sociology to disaster research. Milan, pp. 57-69.

Milliman, J.W. 1982: Modelling regional economic impacts of earthquakes. In: Jones, B.G./Tomazevic, M. (eds.): Social and economic aspects of earthquakes. Proceedings of the 3rd international conference, in Bled, Yugoslavia, 29.6.-2.7.1981, Ljubljana, Ithaca/New York, pp. 175-186.

Mitchell, J.K. et al. 1989: A Contextual Model of Natural Hazards. In: The Geographical Review, vol. 79, pp. 391-409.

Mitchell, W.A. and Barnes, C.T.: Change after an earthquake disaster in Western Anatolia. United States Air Force Academy, Colorado, Department of Economics, Geography and Management, Colorado Springs, Colorado, Januar 1978.

Mitchell, W.A. and Miner, T.H.: Environment, disaster and recovery: A longitudinal study of the 1970 Gediz earthquake in Western Turkey. United States Air Force Academy, Colorado, Department of Economics, Geography and Management, Colorado Springs, Colorado, November 1978.

Music, V.B. 1982: Spatial and urban planning and development in earthquake-prone areas. In: Jones, B.G. and Tomazevic, M. (eds.): Social and economic aspects of earthquakes. Proceedings of the 3rd international conference, in Bled, Yugoslavia, 29.6.-2.7.1981, Ljubljana, Ithaca/ New York, pp. 299-306.

Nimis, P.G. 1978: Friuli dopo il terremoto. Gemona, Artegna, Magnano: Fisica e metafisica di una ricostruzione. Venezia.

Norsa, A. 1979: The reconstruction of Friuli - emergency versus long-term planning. In: Disaster, Vol. 3, Nr. 3, pp. 264-265.

Norton, R. 1980: Disasters and settlements. In: Disasters, Vol. 4, Nr. 3, pp. 339-347.

Okrent, D. 1980 Comment on societal risk. In: Science 208, pp. 372-375.

O'Riordan, T. 1974: The New Zealand natural hazards insurance scheme: application to North America. In: White, G.Fed., Natural Hazards: Local, National Global. Oxford University Press, New York, pp. 217-219.

Palm, R.J. 1990: Natural Hazards. An Integrative Framework for Research and Planning. Baltimore.

Pascolini, M. 1981: Il terremoto e la percezione del rischio sismico.

Petak, W.J., Atkisson, A.A. 1982: Natural hazard risk assessment and public policy. Anticipating the unexpected. New York/Heidelberg/Berlin.

Platt, R.H. 1982: The Jackson Flood of 1979. A public policy disaster. In: Journal of the American Planning Association, Vol. 48, Nr. 2, pp. 219-231.

Qingkang, Y. 1982: Planning for human settlements in disaster-prone areas: The Chinese experience. In: Disasters, Vol, 6, Nr. 3, pp. 202-206.

Quarantelli, E.L. (ed.) 1978: Disasters: Theory and research. Beverly, Hills, Calif.

Quarantelli, E.L. 1982: General and particular observations on sheltering and housing in American disasters. In: Disasters, Vol. 6, Nr. 4, pp. 277-281.

Quarantelli, E.L. 1982: What is a disaster? An agent specific or an all disaster spectrum approach to socio-behavioral aspects of earthquakes? In: Jones, B.G./Tomazevic, M. (eds.): Social and economic aspects of earthquakes. Proceedings of the 3rd international conference, in Bled, Yugoslavia, 29.6.-2.7.1981, Ljubljana, Ithaca/New York, pp. 453-478.

Ronza, R. 1976: Friuli dalle tende al deserto? Scena e retroscena di una ricostruzione mancata. Milano.

Rubin, C. et al. 1985: Community recovery from a major natural disaster. Monograph Nr. 41, University of Colorado, Boulder/Colorado.

Saarinen, T.F.; Sell, J.L. and Husband, E. 1982: Environmental Perception: International Efforts. In: Progress in Human Geography, H.6, pp. 515-546.

Saraceno, E. 1981: Emigrazione e rientri. Il Friuli-Venezia Giulia nel secondo dopoguerra, Udine.

Schenkel, F.: I meccanismi di controllo del finanziamento pubblico nel mercato edilizio della ricostruzione. In: Fabbro, S. (ed.): 1976-1986 La ricostruzione del Friuli. Udine 1986, S. 146-152 bzw. p. 169.

Seaman, J. 1980: The effects of disaster on health: A summary. In: Disasters, Vol. 4, Nr. 1, pp. 14-18.

Senn, L. 1982: Il ruolo delle imprese nelle politiche regionali per la promozione industriale. In: Scritti in onore de J. Gasparini. Milano, pp. 1011-1034.

Slovic, P. et al., 1979: Rating the Risks. In: Environmental, Vol.21, Nr.3, April, pp.14-39.

Smith, N. 1989: Geography as Museum: Private History and Conservative Idealism. In: Entrikin, J.N. and S.D. Brunn (eds.): Reflection on Richard Hartshorne's "The Nature of Geography". Occasional Publications of the Association of American Geographers. Washington, pp. 91-120.

Snarr, D.N. and Brown, E.L. 1982: Attrition and housing improvements: A study of post-disaster housing after three years. In: Disasters, Vol. 6, Nr. 2, pp. 125-131.

Snarr, D.N., Brown, E.L. 1980: User satisfaction with permanent post-disaster housing: Two years after Hurricane Fifi in Honduras. In: Disasters, Vol. 4, Nr. 1, pp. 83-91.

Di Sopra, L. 1986: Impact Magnitude. Theoretical Models. The Friuli Experience Synthesis. Milan.

Di Sopra, L. and Pelanda, C. (eds.) 1984: Teoria della vulnerabilitá. Introduzione multidisciplinare. Milano.

Di Sopra L., and Schiavo, F. 1983: First hypotheses for the construction of vulnerability tables for persons involved in earthquakes in Italy. Medical aspects. Preliminary Paper Nr. 83-4, Gorizia (ISIG).

Stephens, L.H. and .J. Green (eds.) 1979: Disaster Assistance. Appraisal, Reform and New Approaches. Baltimore.

Steuer, M.: Wahrnehmung und Bewertung von Naturrisiken am Beispiel zweier ausgewählter Gemeindefraktionen im Friaul. Münchener Geographische Hefte 43, Kallmünz/Regensburg 1979.

Stagl, R.: Terremoto e ricostruzione secondo gli uffici tecnici dei 45 comuni disastrati. In: Ricostruire, Anno quinto, No.15, pp. 8-19.

Strassoldo, R. and Cattarinussi, B. (eds.) 1978: “Friuli: la Prova del Terremoto”, Milano.

Stratta, J.L. and Wyllie, L.A. Jr.: Friuli, Italy eathquakes of 1976. EERI (Earthquake Engineering Research Institute), Berkley, Calif., 1979.

Stretton, A. 1976: The Furious Days. The Relief of Darwin. Sydney, London.

Stretton, A. 1979: Ten Lessons from the Darwin Disaster. In: Heathcote and Thom (eds.): Natural Hazards in Australia. Australian Academy of Science. Canberra, pp. 503-507.

Stukelj, P. 1982: Rescue operations after an earthquake. In: Jones, B.G./ Tomazevic, M. (eds.): Social and economic aspects of earthquakes. Proceedings of the 3rd international conference, in Bled, Yugoslavia, 29.6.-2.7.1981, Ljubljana, Ithaca/New York, pp. 405-412.

Sutphen, S. 1983: Lake Elsinore disaster: The slings and arrows of outrageous fortune. In: Disasters, Vol. 7, Nr. 3, pp. 194-201.

Turoldo, D.M. (ed.) 1977: Friuli, un popolo tra le macerie, Roma.

Turnsek, V. 1982: Earthquakes as a social problem. In: Jones, B.G./Tomazevic, M. (eds.): Social and economic aspects of earthquakes. Proceedings of the 3rd international conference, in Bled, Yugoslavia, 29.6.-2.7.1981, Ljubljana, Ithaca/New York, pp. 133-146.

Valussi, G. 1971: La populazione del Friuli-Venezia Giulia. In: Istituto per l'Enciclopedia del Friuli-Venezia Giulia: Enciclopedia Monografica del Friuli-Venezia Giulia, part I, Il Paese, vol. II, pp. 759-805.

Valussi, G. 1972: Il fenomeno migratorio in Friuli fra processi di deruralizzazione e industrializzazione. In: Atti del 1° Convegno ifres, Udine, pp. 104-129.

Valussi, G. 1973: Le direttrici dello sviluppo economico nel Friuli-Venezia Giulia. In: Atti della Tavola Rotonda su Poli, Assi e Aree di sviluppo economico con particolare riguardo alle regioni sottosviluppate (Roma 1972). Società Geografica Italiana, Roma, pp. 391-405.

Valussi, G. 1974: L'emigrazione nel Friuli-Venezia Giulia. In: Istituto per l'Enciclopedia del Friuli-Venezia Giulia: Enciclopedia Monografica del Friuli-Venezia Giulia, part II, La vita economica, vol. II, Udine, pp. 853-928.

Valussi, G. 1977: Il Friuli di fronte alla ricostruzione. In: Rivista Geografica Italiana, LXXXIV, pp. 113-128.

Valussi, G. 1978: La mobilità della popolazione friulana dopo gli eventi sismici del 1976. In: Atti del Convegno di studi sui fenomeni migratori in Italia, Università di Udine, Istituto di Geografia della Facoltà di Lingue e Letterature Straniere, Udine, pp. 383-409.

Valussi, G. 1979: La mobilità della popolazione. In: AA.VV., Friuli 1976. Una ricerca socio-economica su sei Comuni dell'area terremotata, Confederazione Generale dell'Industria Italiana, Federazione Regionale degli Industriali del Friuli-Venezia Giulia, Roma.

Ventura, F. 1982: The earthquake in Campania and Basilicata (Italy) of 23rd November 1980 - The nature of the state's emergency interventions and the future quality of reconstruction. ISIG Publication Nr. 82-5, Gorizia.

Verney, P. 1979: The earthquake handbook. London.

Volpato, G. 1978: Economia e strategia delle innovazioni. In: Saraceno, P. (ed.) Economia e direzione dell'impresa industriale. ISEDI, Milano, pp. 197-257.

Walters, K.J. 1978: The reconstruction of Darwin after cyclone Tracy. In: Disasters, Vol. 2, Nr. 1, pp. 59-68.

Western, J.S. and G. Milne1979: Some Social Effects of a Natural Hazard: Darwin Residents and Cyclone Tracy. In: Heathcote and Thom (eds.): Natural Hazards in Australia. Australian Academy of Science. Canberra, pp. 488-502.

Westgate, K.N. 1979: Land-use planning, vulnerability and the low-income dwelling. In: Disasters, Vol. 3, Nr. 3, pp. 244-248.
Wolensky, R.P. 1985: Power, policy and disaster: The political organization impact of a major flood. University of Wisconsin, Stephens Point, Wisconsin.
Wright, J.D. et al. 1979: After the clean-up. Long range effects of natural disasters. Beverly Hills, Calif.
Zamberletti, G. 1986: Lessons to be learned from earthquakes. In: International Civil Defence, Nr. 370/371, Geneva, April/May pp.1-4.

Index

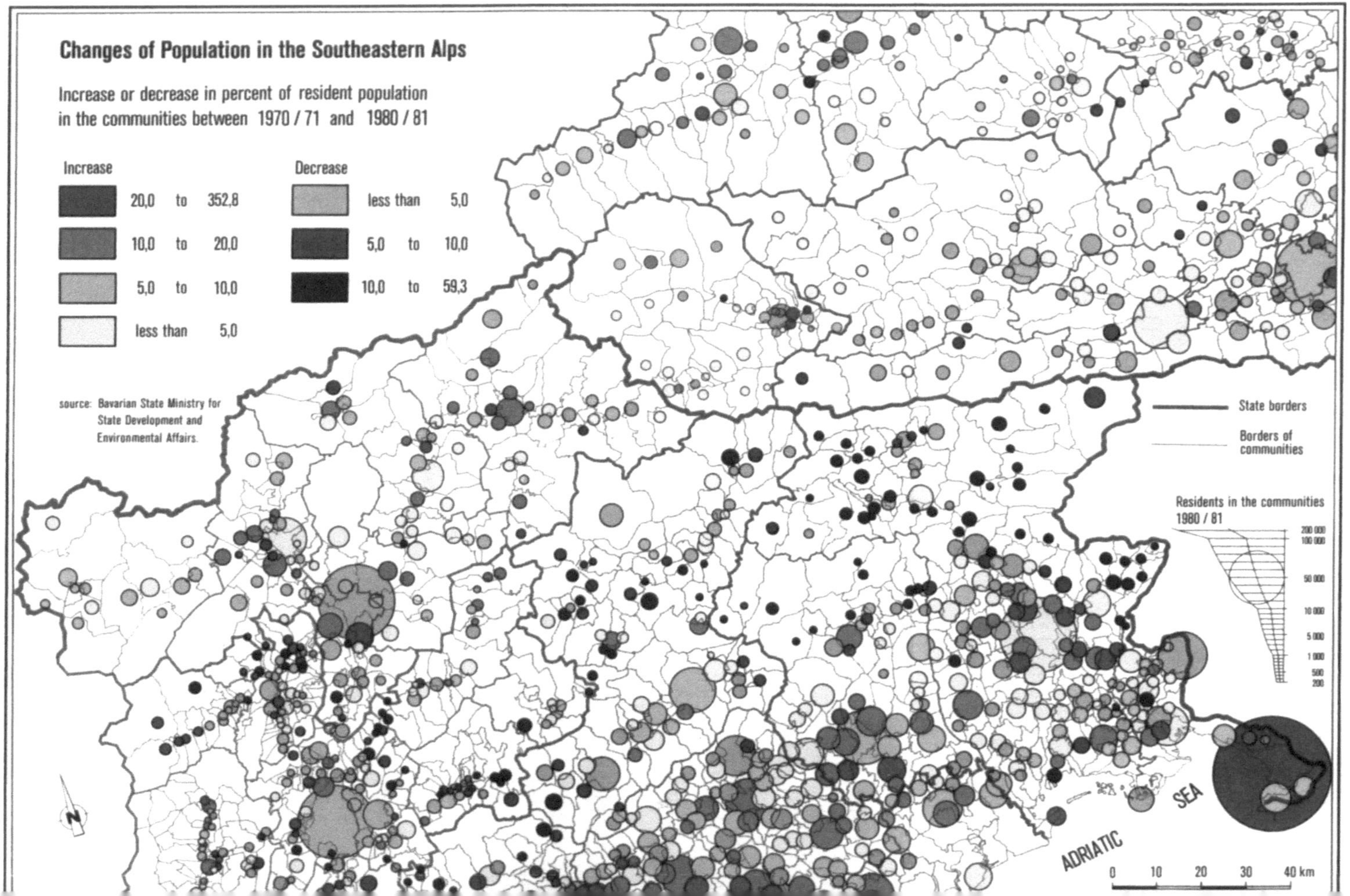
Changes of Population in the Southeastern Alps
Increase or decrease in percent of resident population
in the communities between 1970 / 71 and 1980 / 81
Increase
20,0 to 352,8
10,0 to 20,0
5,0 to 10,0
less than 5,0
Decrease
less than 5,0
5,0 to 10,0
10,0 to 59,3
source: Bavarian State Ministry for State Development and Environmental Affairs.
State borders
Borders of communities
Residents in the communities 1980 / 81
200 000
100 000
50 000
10 000
5 000
1 000
500
200
ADRIATIC
SEA
0 10 20 30 40 km
N

MIX
Papier aus verantwortungsvollen Quellen
Paper from responsible sources
FSC® C105338

If you have any concerns about our products,
you can contact us on
ProductSafety@springernature.com

In case Publisher is established outside the EU,
the EU authorized representative is:
Springer Nature Customer Service Center GmbH
Europaplatz 3, 69115 Heidelberg, Germany

Printed by Libri Plureos GmbH
in Hamburg, Germany